Remote Sensing for Landscape Ecology

Monitoring, Modeling, and Assessment of Ecosystems

SECOND EDITION

Remote Sensing for Landscape Ecology

Monitoring, Modeling, and Assessment of Ecosystems

SECOND EDITION

Ricardo D. Lopez
Robert C. Frohn

CRC Press
Taylor & Francis Group
Boca Raton London New York

CRC Press is an imprint of the
Taylor & Francis Group, an **informa** business

CRC Press
Taylor & Francis Group
6000 Broken Sound Parkway NW, Suite 300
Boca Raton, FL 33487-2742

Printed and bound in India by Replika Press Pvt. Ltd.

Printed on acid-free paper

International Standard Book Number-13: 978-1-4987-5436-1 (Hardback)

Library of Congress Cataloging-in-Publication Data

Names: Lopez, Ricardo D., author. | Frohn, Robert C. Remote sensing for landscape ecology.
Title: Remote sensing for landscape ecology : monitoring, modeling, and assessment of ecosystems / Ricardo D. Lopez.
Description: Second edition. | Boca Raton : CRC Press, [2017] |
Previous edition: Remote sensing for landscape ecology :
new metric indicators for monitoring, modeling, and assessment of ecosystems / Robert C. Frohn (Boca Raton : Lewis Publishers, 1998). |
Includes bibliographical references and index.
Identifiers: LCCN 2017006901 | ISBN 9781498754361 (Hardback : acid-free paper)
Subjects: LCSH: Landscape ecology--Remote sensing. | Geographic information systems.
Classification: LCC QH541.15.L35 F76 2017 | DDC 577.5/50285--dc23
LC record available at https://lccn.loc.gov/2017006901

Visit the Taylor & Francis Web site at
http://www.taylorandfrancis.com

and the CRC Press Web site at
http://www.crcpress.com

This book is dedicated to the influential people in my life, especially to my wife

Debra, for her unwavering energy, positivity, and love; to my mother Lynn,

who always encouraged my imagination and creativity; and to my father

Ricardo Carlos, who demonstrated how enjoying one's work leads to lifelong

fulfillment. I also dedicate this work to the many hardworking employees

of the United States Environmental Protection Agency for their persistent

pursuit of sound science, environmental justice, and public service.

Ricardo D. Lopez

To the memory of Robert C. Frohn, author of the first

edition. This book is your continuing legacy.

Contents

Preface

The broad goal of this book builds upon the first edition by elucidating and demonstrating landscape metrics and indicators that show responses to characteristic variations in remotely sensed ecological information. To achieve this goal, it is necessary to go beyond the first edition of this book and meet the multiple objectives that are a typical part of all landscape ecological projects. These are, in the twenty-first century, a characteristic expectation and need of society, and thus decision makers in a variety of realms. In addition to developing landscape metrics that are sensitive to change in patch characteristics, such as configuration and complexity, a thorough understanding of the environmental gradients of change that are of relevance to users' decision-making needs or goals must precede the design of a project. The development of landscape metrics and indicators that are predictable with regard to changes in the biophysical characteristics of the environment as well as the spatial and temporal resolution of data used are among the key objectives of this book. To accomplish these goals, the linkages between ecosystem functions of interest and information provided by remote sensing data, as well as any limitations of either of these elements, are necessary to better understand in order to establish ecological significance, statistical significance, and relevance to communities and decision makers.

The metrics and indicators developed in this book are practical in nature, as well as quantitative; a major addition to this edition of the book is the practicality of the approaches taken, with specific examples and outcomes. Although the first edition advocated powerfully for the use of specific metrics for theoretical reasons, which are sound and very important to consider, there is an emerging paradigm within community planning, natural resource management, ecological restoration, and conservation as a whole that demands the inclusion of landscape scale analyses within all societal development, which has become well established and institutional in some cases during the past two decades and since the completion of the first edition. The metrics outlined in this edition are based upon these societal demands, and utilize a balance of scientific and geostatistical approaches necessary to ensure fidelity under a number circumstances, driven by geography, biophysical conditions, and data availability. All of the landscape metrics and indicators described in this edition are provided as diverse examples for application and modification, all founded in sound science as well as the user's specific decision-making needs or goals.

This edition goes far beyond applying theory to the four project case studies by addressing a broad range of challenging project needs encountered in today's sophisticated atmosphere of melded scientific and societal approaches for broad-scale landscape ecology, which naturally impinge upon both the spatial and temporal gradients of all of the regions described. Briefly introducing the needs addressed and the geographic areas covered by this edition are the landscape ecological metrics and use of diverse outputs of remote sensing in the entire Laurentian Great Lakes (United States and Canada) Watershed; landscape ecological metrics and indicators of water quality derived from both remote sensing and field information in the Ozark Mountains (Missouri and Arkansas) Watershed; a landscape ecological focus on sea level rise in both coastal California and coastal North Carolina areas as it encroaches on the landscape in the coming century; and landscape hydrologic conditions and modeling in the Missouri River and Mississippi River Watersheds. All four geographies presented in this edition utilize well-established landscape metric fundamentals and include novel methods for practical application, which demonstrate the capability of a host of metrics for distinguishing among land cover types and landscape changes.

The book provides important new outcomes from research and development on the integration of remote sensing, geographic information systems, and landscape ecology metrics for modeling, monitoring, and assessment of ecosystems at a variety of scales. The long-term value of such research and development of applications has been incompletely realized; however, as society continues to employ these technologies into the future and methods continue to be developed to improve understanding of the world in which we live, a better awareness of "systems thinking," especially for developing predictive models, can be utilized to shape decisions, and those decisions can be realized in demonstrable ways that serve society. Because the book is concerned with the development and application of landscape ecology metrics, it may be misunderstood that the metrics and examples used are exclusive of other work in this extremely large field of landscape ecology; this is not the case. The work described in this edition is meant to address a wide spectrum of similar work, and it is not the authors' intention to suggest that other studies are without merit or not demonstrative in their own right, but rather that the examples provided in this edition are emblematic of the precepts and concepts embodied within the current discipline of applied landscape ecology. The authors are indebted to the researchers of past studies in landscape ecology, especially those who have been devoted to the development of the numerous array of landscape ecology metrics and indicators as a whole.

Acknowledgments

This book integrates and synthesizes a complex topic, utilizing a host of methods and approaches for utilizing remote sensing and other information for landscape ecological analyses. As such, a number of fine people played key roles in the development of these techniques, including Lee Bice, Don Ebert, Curt Edmonds, Ed Evanson, Brenda Groskinsky, Michael Jackson, Maliha Nash, Kamal Qaiser, Caroline Torkildson, Rick Van Remortel, Larry Woods, and Yongping Yuan. For their hard work and skills, the author is forever grateful.

Author

Ricardo "Ric" D. Lopez, PhD, has been a leader in the field of landscape ecology over the past three decades. During his tenure in academia and public service with the Ohio Environmental Protection Agency, the U.S. Environmental Protection Agency, and the U.S. Forest Service, he has led in the geographically diverse applications of remote sensing and field-based approaches for geospatial analyses, as applied to both theoretical and applied environmental topics. This body of work includes the monitoring and assessing of terrestrial, aquatic, and transitional ecosystems; invasive plant species; multi-scale indicators of sustainability; and solutions to risk-based landscape ecology issues. A native of California, Dr. Lopez has spent much of his life exploring, appreciating, and writing about the diverse aspects of complex landscapes, from the tropics to temperate regions, bringing his expertise as a landscape ecologist to bear on specific local, regional, and global environmental issues. He earned a BS in Ecology, Behavior, and Evolution at the University of California, San Diego, and master's and doctoral degrees in Environmental Science at The Ohio State University, with an emphasis in landscape ecology and wetland ecology. Dr. Lopez is currently the Director of the U.S. Forest Service's Pacific Southwest Research Station—Institute of Pacific Islands Forestry in Hilo, Hawai'i.

1

Introduction

1.1 New Challenges for the Landscape Sciences

This book expands on the first edition (Frohn 1997) by focusing on the specific applications of the fragmentation and patch complexity metrics that are often utilized in landscape ecology, as derived from remote sensing data. To that end, the reader will find that their project work shares a number of similarities with the examples in this book, albeit perhaps in different geographies or other circumstances. Those "other projects" that are familiar to the reader will be informed by the processes used and the outcomes generated from the presented projects in this book, which integrate well with a large number of disciplines (from biology to engineering to hydrology), all of which collectively define the collective of disciplines (and individuals) who are the members of an excellent and diverse landscape sciences team.

Numerous quantitative measurements of landscape pattern have been used in the field of landscape ecology (e.g., as outlined by Krummel et al. 1987 and O'Neill et al. 1988) several decades ago, which has served as a basis for quite a bit of landscape ecological work since then. These measurements, often called metrics or indicators, have been used to link ecological and environmental processes with patterns found within the larger geographic matrix, commonly referred to as the *landscape* (Forman 1995). The applied use of metrics often requires the practical use of existing data and these fundamental metrics are often simplified or modified from three fundamental metric types (i.e., dominance, contagion, and fractal dimension), and derivations thereof, to match the biophysical environmental conditions, ecosystem processes and characteristics, or the decision-making goals of the users in a particular geography. The specific application of numerous landscape metrics and indicators are applied and discussed in detail in Chapters 4 and 5.

Various landscape ecology metrics have been used to quantify aspects of spatial patterns, and to correlate them with actual ecological processes (e.g., O'Neill et al. 1988; Turner and Gardner 1991; Baker and Cai 1992; McGarigal and Marks et al. 1994; Riitters et al. 1995). In particular, spatial pattern metrics that are related to contagion and dominance (based on information theory) and fractal dimension (based on fractal geometry) have been used extensively throughout the landscape ecology community (e.g., Krummel et al. 1987; O'Neill et al. 1988; Turner and Gardner 1991; Milne 1991; Wickham and Riitters 1995), and indeed serve as the basis of a host of developed metrics since these earlier days of contemporary landscape sciences. These three metric types have been extensively utilized in various modifications for implementation of watershed integrity indicators; landscape stability and resilience indicators; and biotic integrity and diversity indicators, as demonstrated in the latter chapters of this book. Several tests of these metrics and indicators have also occurred within the remote sensing community (e.g., Wickham and Riitters 1995; Wickham et al. 1996; O'Neill et al. 1996; Fenga and Liu 2015).

Dominance metrics have been used as a landscape diversity measure by determining the equality of the proportion of land cover types across a particular landscape (O'Neill et al. 1988). High dominance indicates that one or more land cover types are covering the landscape. Low dominance indicates that land cover types have nearly equal proportions. However, dominance does not necessarily indicate diversity of the landscape. For example, a landscape with two land cover types with 50% proportion will have the same dominance value as one with 10 land cover types with 10% proportion. Thus, the same arguments that apply to species diversity indexes apply to the dominance landscape metric of diversity. Also, the dominance metric does not actually give a quantitative measurement of landscape pattern, although it has been referred to as a spatial metric. This book demonstrates the use of the dominance landscape metrics (e.g., the use of the Shannon–Wiener Index and Simpson's Index) within this context and understanding of the limitations, and within the context of other ecological metrics, such as contagion and fractal dimension metrics—the combined application of these three types of metrics, and an understanding of their uses and limitations, leads to the concept of ecological significance of landscape metrics.

Contagion metrics have been used in ecosystem analyses to quantify the amount of clumping or fragmentation of patches on a landscape (O'Neill et al. 1988). They have been utilized to relate the effects of contagion patterns on ecosystem processes such as habitat fragmentation, vegetation dispersal, and animal movements (e.g., Turner and Ruscher 1988; Turner 1989, 1990a, 1990b; Graham et al. 1991; Gustafson and Parker 1992; Li and Reynolds 1993; USEPA 1994, 1996). Fractal dimension metrics have been used in ecosystem analysis to quantify the complexity of patch shapes on a landscape (e.g., Krummel et al. 1987; O'Neill et al. 1988; De Cola 1989; Lam 1990); all of these metric categories have also been used to measure the degree of human disturbance on the landscape. The underlying theory of these metrics is that natural boundaries, such as those for vegetation, have relatively more complex shapes than those that are a result of human activity, such as agricultural fields. As human disturbance increases, the fractal dimension of the landscape decreases (e.g., Krummel et al. 1987; O'Neill et al. 1988; Turner and Ruscher 1988; De Cola 1989). The contagion and fractal dimension metrics have been evaluated for their sensitivity to variations in remote sensing data and raster data structures (e.g., Kalkhan 2007) and findings are that orientation, shape, and resolution of remote sensing data elements can lead to more or less usable information, relative to ecological processes on the ground. These metrics were originally developed to focus attention toward quantification of landscape pattern and to encourage the development and application of new or improved metrics in ecosystem analysis (O'Neill 1996; personal communication in Frohn 1997).

The utility of any landscape metric is dependent on its maintaining a consistent response to observed phenomena. This does not occur when the fundamental assumptions applied in its formulation are violated. Even in cases where the use of a contagion or a fractal dimension metric may be appropriate, there are a number of characteristics that affect the quality of map and image data, including spatial resolution, geometric registration, and level of classification. In order for a landscape metric to be effective it should also be relatively insensitive to arbitrary sampling characteristics while being very sensitive to the specific spatial patterns. Since remote sensing and other landscape data are captured in a wide variety of geometric representations, landscape metrics must be formulated to compensate for specific sampling geometries in order to facilitate comparison and integration across scales and among different studies.

In addition to technical remote sensing challenges, a critical additional consideration has emerged in the past decade that is relevant to contemporary and future applications of remote sensing for landscape ecology. These considerations drive not only the ultimate use of the outputs but also consideration of accuracy and precision in data incorporation of the social dimensions of the landscape. Evidence of these changes are the incorporation of the terms *sustainability* and *ecosystem services* into the decision-making goals and objectives of many landscape ecology projects and research, particularly in the past decade. The concepts of sustainability are not new; however, their current incarnation that unifies social and ecological perspectives of nature provides a challenging new goal for all landscape ecologists to quantify ecosystems (e.g., identification and characterization), their condition (e.g., ecological functions), and their relationship(s) with society (e.g., ecosystem services) (Reid et al. 2010). From the most general concept of sustainability comes a broad view of environmental and ecosystem management issues, which offer approaches for going beyond solely the technological solutions to environmental problems by integrating social participation and policy dialogue with ecological inventorying, monitoring, and assessment activities.

Among the many emerging critical and future threats to the sustainability of ecosystems are soil loss and degradation; water scarcity; and the loss of biological diversity (Running et al. 2004), regardless of the sociological contexts. The perceptions of these environmental problems vary tremendously, depending upon a number of socioecological factors. If remote sensing data, and indeed all associated geospatial information, is to have an impact on the users in these areas, the information produced needs to be compelling, accurate, and easily accessible to the user (i.e., must have high impact and availability). Some argue for an approach that addresses this complexity as a "multilevel stakeholder approach to sustainable land management," for finding feasible, acceptable, viable, and ecologically sound solutions at local scales. As explored in-depth in Chapter 6, a number of international programs and bilateral cooperation projects have taken this perspective and started using a sustainable land management approach, such as in the case of the United Nations Capital Development Fund. A sustainability paradigm such as sustainable land management requires that a technology follow some fundamental principles: (1) ecological protection, (2) social acceptance, (3) economic productivity, (4) economic viability, and (5) risk reduction (UN 2012). Accordingly, a technological approach to resource management that is sustainable would have to be developed using criteria for a particular and locally relevant land use, and would likely not be applicable everywhere. This method encourages the full exploration and inclusion of the economic, social, institutional, political, and ecological dimensions of the community/ geography in question. Global environmental professionals have suggested the efficacy of this approach of tying environmental science, technology, and society by explicitly linking research on global environmental change with sustainable development (Reid et al. 2010), which would necessitate an increased use of remote sensing and geospatial analysis for monitoring ecosystem conditions, and also for measuring feedback loops between environmental conditions and societal values and activities.

1.2 Goals and Objectives of This Book

The principal goals of this book are to thoroughly outline the various advantages and limitations of utilizing remote sensing data for landscape ecology, and to provide practical examples of the use of landscape metrics within the context of the aforementioned advantages and challenges with a full discussion of the many considerations that must be made when selecting remote sensing data types, exploring project designs, and utilizing the outputs of projects. In the first edition of this book, the author approached this challenge from a purely quantitative perspective, focusing mainly on the specific methods that would ensure greater certainty in the technical/theoretical applications of remote sensing data. This book expands this approach by soberly looking at the current state of the art, in terms of the remote sensing data utilized for developing landscape metrics and indicators as utilized in the field, tying the technological and scientific elements to practical management goals and perspectives, which drive today's needs in society. Specific examples go far beyond the theoretical and utilize remote sensing data in a variety of key geographies, with user needs that must be met, and have been specifically selected to demonstrate techniques for expanding and building upon the earlier edition's information. Each of the examples given are actively used by end users and decision makers, making them an excellent demonstration of both the theoretical and applied uses of remote sensing and landscape ecology. This approach is not intended to completely overcome the uncertainties described above in the challenges of Section 1.1, rather they are meant to demonstrate the approaches for balancing uncertainty with project needs and professional approaches, which the reader is no doubt likely to experience in present or future circumstances. In this book, three specific concepts and approaches are used to overcome these challenges, which were outlined previously:

1. The concept of ecological significance is key to understanding trends and multiple streams of data with ecological theory as a basis of understanding those trends and data.
2. The concept and approach of trajectory analysis is important for interpreting multiple streams of data, including both the qualitative and quantitative, leading to further inquiry and discovery of trends.
3. The concept and approach of hypothesis generation (without a fear of endless analysis) in a particular geography or ecosystem type, which enables the use of data that is imperfect, within a controlled analytical environment, where assumptions are stated and utilized to guide user's interpretation, decision making, and further generation of hypotheses and focused work (perhaps at another scale or a subsetted area of analyses).

This book also utilizes these three complementary concepts/approaches, which were initially explored in the first edition, building further upon what has been learned in the past two decades of landscape ecological applications. The expansive approach taken in this edition recognizes the tremendous progress made since the first edition, and the need to continue that progress in the discipline of landscape ecology as a whole, despite substantial uncertainty that is intrinsically a part of environmental data collected from a distance, i.e., remote sensing of the environment, while providing some fundamental concepts and approaches for utilizing these constantly improving sources of data. A much more expansive approach in this edition provides the reader with additional tools, techniques, and perspectives to comparatively explore the approaches taken under the challenging circumstances of uncertainty to further the goals of landscape ecologists and remote sensing scientists/practitioners, specifically as it pertains to landscape monitoring, landscape modeling, and the assessment of ecosystem functions.

In general, the goals and objectives of this book are met by carrying out a series of tasks. These tasks include the following:

- A contemporary overview of landscape ecology metrics for the monitoring and assessment of landscape change through application and an improved understanding of ecosystems from that perspective.

- A contemporary evaluation of landscape metric types (dominance, contagion, and fractal dimension) on a conceptual basis, and through specific applications.

- A comparison of landscape metrics using real-world data, conditions, and circumstances for various gradients of change, including both the biophysical and the societal considerations of the landscapes involved.

- An analysis of the strengths, weaknesses, opportunities, and limitations for utilizing each of the metrics explored, and the implications for future work in the field of remote sensing for landscape ecology.

1.3 Utilizing This Book to Its Fullest Benefit

This book is intentionally written for the full diversity of environmental and resource professionals at the broad-level and sub-fields of engineering, ecology, resource management, climate sciences, and policy development at the local, state, regional, national, and global scales. This book is certainly useful as a reference and handbook for readers increasing their understanding of broad scale landscape ecology work that involves the direct use of remote sensing data, and also provides a wide variety of geospatial data outputs, all of which provide a full range of examples to guide readers' project work. Results and information in the several applied examples in this book contain some key nuggets of information that come from experience and application, many of which address the important linkages that are not always apparent in every project, spanning resource parameters, a variety of scales, and decision making processes. Such use of technology and science to serve the needs of broad scale issues can inform national and international policies related to, for example, restoration; clean water; climate change; human safety, health, and well-being; and sustainable development (Doyle and Drew 2008), with direct relevance to international commissions, conventions, protocols, and agreements that require compiling information from community- and place-based analyses, allowing for broad scale application of the relevant elements from fine-scale needs and decisions. A worthy example, discussed in detail in Chapter 3 with applications in Chapter 4, is the ongoing work in the transboundary ecosystem of the Laurentian Great Lakes, where Canada and the United States have successfully collaborated to monitor and assess the ecological functions and services of the entire Great Lakes Ecosystem, known as the Great Lakes Restoration Initiative, with tremendous strides in recent and novel research and restoration. The initiative was initially funded in 2010 and is now reaching nearly $2 billion through the Fiscal Year 2015 President's Budget. This particular example provides a specific vision for future successful linkages between the applied work of remote sensing specialists and landscape ecologists, with an in-depth presentation of this specific transboundary solution to evaluating and monitoring massive landscapes; in this case, the entirety of the Great Lakes Watersheds. This in-depth example is followed by several other important landscape projects with additional intricacies and subtleties of analysis, which similarly link fine-scale information and needs to broad-scale understanding of environmental conditions, each serving decision makers and communities well.

This book uses the selected applied examples to highlight current remote sensing, geographic information systems, geostatistical, and modeling techniques to address the challenges you may encounter now, or in the future, in your particular geography or professional circumstances. The techniques demonstrated in this book are selected to show the breadth of applications for a diversity of landscapes and ecosystem types, resource conditions, and societal dimensions that are encountered every day by practitioners.

Throughout the book, a number of Internet sites and other resources from the literature are noted for reference, which equip you as the reader with a tremendous amount of integrated information that provides you with time-saving and clear pathways toward the solution for specific challenges that you may encounter when utilizing remote sensing for landscape ecology. All of the methods and approaches outlined in this book provide both new users and seasoned professionals with practical tools for success in the ever-changing world of landscape sciences, which now requires, at least, a reasonable facility with possibilities for the use of remote sensing to communicate effectively with like-minded professions. The complexity of present-day situations presented to ecologists, particularly if the focus is beyond plot-based work, requires an agility of understanding, conceptualization/articulation of capabilities, design skills, and the knowledge of the steps needed to implement a landscape scale project (or at least an ability to direct others on a viable path), by providing the necessary detail and synthesis of numerous available/potential approaches. This book was designed to provide several quality examples of successful and complex landscape ecology projects (Chapters 4 and 5), all of which have a number of the complexities mentioned above, so as to explain how you can develop a fully successful project with substantive outputs and outcomes under such circumstances. The approaches and methods described in this book should be read and analyzed by you, the reader, in such a way as to discover and extract similarities in your work, so as to recognize and translate the circumstances in the work described in this book into your own application of the same or similar technology, science, and societal parameters. To aid in this utilization for the contemporary applications of the reader, several new aspects are included in this edition of the book.

1. A practical update of remote sensing data types and geospatial methods
2. Demonstrations of specific examples, which are project driven
3. Descriptions of any pitfalls of using ecological data at landscape scales, with solutions
4. Discernment of alternative techniques for a variety of practitioners
5. Inclusion of specific linkages between field-based and landscape-based remote sensing and ecological practices
6. Updated resources for practitioners

1.4 Significance of Landscape Ecology Research

It is important to realize, initially, that there is a theoretical basis for the two rich disciplines of landscape ecology and remote sensing, as well as the many allied sciences involved in the work we will discuss in this book. Ultimately, landscape metrics are employed to create quantitative measures of commonly observed spatial patterns found on a map or within a remote sensing image or data set. Just take a look at a map or image derived from remote sensing imagery, or within a data set associated with the map or image, and notice the multitude of patterns illuminated. If making a visual assessment of the map or an image, perhaps the landscape is full of rectangular geometric shapes, which may be indicative of agricultural fields. Or perhaps one finds an image that contains an array of adjacent circles, indicative of fields with center point irrigation systems. In other areas one may see a regular grid of intersecting lines, such as those often found in residential areas. One may notice in a scene that there are many elongated complex shapes such as those found in association with geologic formations. Or, there may be a series of thin elongated narrow strips with a more regular shape, indicative of geologically "folded" mountain ridges and valleys. One notices these similar patterns across the world, and indeed, all of these patterns are observable in each of the focal geographies described in later chapters, within the Great Lakes Basin, in the marine coastal areas of California and North Carolina, in the rolling Ozark Mountains of Missouri and Arkansas, and in the Midwestern large river systems of the United States. These commonly observed patterns help us as landscape ecologists understand common processes that may occur in different geographies without the need to completely "reinvent the wheel" in each new study area. Nonetheless, hypothesis testing is always needed to confirm that the characterization of the landscape is accurate, and that metrics are an accurate measure of ecological processes on the ground.

One may also notice the particular arrangement of patches across a landscape. Perhaps patches are fragmented into thousands of small forested wetland patches along a river bank, such as in the Midwestern Missouri River Basin. Or they may exist as large patches of forest with cleared areas where development has occurred, such as in the Ozark Mountains. Basically, when one views a satellite image she may notice or identify many elements that, when combined, characterize the physical aspects of the scene. These elements include tone or color, shadow, illumination, location, association, objects, and process. In addition, the observer may notice varying shapes and sizes of those elements, textures, and patterns. It is the quantification of these groups of elements into a measurable variable that creates what we refer to as a *landscape metric*. Thus, all of the patterns mentioned previously can be quantified and distinguished from one another through the use of landscape metrics, in an infinite variety of combinations. Quantifying these patterns is critically important for a number of reasons, not least of which is that images need to be described by an observer in such a way as to convey comparability and condition, if necessary—this is a communication need for all analysis outputs, but is most important for images and complex geospatial data sets that contain more information than can be humanly possible to process at one time. Consider the complex matrix of geology, soil, and geomorphology (e.g., ridges, sheer cliffs, and valleys) in the Ozark Mountains. How does one describe the shapes, patterns, and textures of this area in a few words? Landscape metrics can potentially quantify these spatial patterns in one or two variables. But, more importantly, the use of landscape metrics by researchers can facilitate the detection of patterns of change that are not readily visible to the human eye nor easily detectable by a human analyst, when utilizing the power of remote sensing and the expertise of remote sensing scientists and practitioners.

Another critical reason why landscape metrics are necessary is to better understand our surroundings in terms of those landscape patterns and ecological and environmental processes, and importantly, so these elements can be linked quantitatively. For these reasons, landscape ecology is indeed the study of the effects of landscape patterns and their changes on ecological processes, and the understanding of these relationships is the constant pursuit of those engaged in this discipline. By quantifying spatial patterns and their changes, landscape ecologists endeavor to quantify their effect on ecological processes, and thus we can study changes in habitat of a particular species or community of organisms and determine whether the habitat has become too fragmented for the species, or an entire biological community, to persist in a particular geographic location. With this fundamental approach understood, we can move forward to determine, for example, the complexity of shapes of a given habitat type (e.g., oak-dominated forests) or determine if certain organisms (e.g., black bear) may travel among locations, given certain impediments between various land cover types and the resources at the locations (i.e., accounting for distance or land cover types). All of these ecological functions, and more, can be theoretically determined and applied from remote sensing data, provided we adequately integrate, test, and apply landscape ecological metrics and remote sensing approaches thoughtfully.

There are a number of aspects of the discipline of geography, particularly remote sensing and geographic information systems, where this book makes a significant contribution. One fundamental contribution is how the combination of both remote sensing and landscape ecology bridges the disciplines of ecology and geography. More specifically, the melding of methodologies and applications of remote sensing and landscape ecology brings knowledge concerning quantitative landscape processes to users of remote sensing and geographic information systems. Of course, remote sensing has long been used as a means for providing data for environmental studies; however, to date, the use of remote sensing technology to characterize landscape patterns and relate those patterns to ecological processes has not been entirely explored, relative to other more integrated disciplines that utilize technology, for example, and analogously, biomedical engineering.

There are a number of other contributions that this book can make to geography, particularly remote sensing and geographic information systems (GIS). First, relevant landscape ecological metrics could certainly be used to improve the classification of remote sensing data in a number of data sets. For instance, some of the landscape ecology work accomplished for this book allowed for the ability to characterize entire watershed basins, coastlines, and instream conditions. This gives notion to what has been referred to as a *spatial signature* of a given land cover type (Frohn 1997). By placing more emphasis on spatial pattern with landscape metrics for land cover classification or combining landscape metrics with spectral or other biological or biophysical information, classification products from remote sensing data can be greatly improved in this regard.

This book is special in that it also strongly emphasizes the practical uses of a number of classical spatial analysis methods within the context of specific societal needs and satisfies the need for practitioners to better determine and infer the biological and biophysical conditions within larger and larger landscapes, worldwide. The authors welcome the reader to consider the similarities, and perhaps the dissimilarities, between the context of the work outlined in this book with their own work context, to aid in developing their plans for metric selection, scale of work, structuring of projects, or other aspects of the outlined projects in Chapters 4 and 5.

1.5 Selection of Study Sites

In this edition, four relatively large extent geographic areas were chosen for the purposes of analysis and demonstration of landscape ecological methodologies and the use of remote sensing to achieve the specific goals of each of the four areas. The four study areas were chosen to assist in these analytical and demonstration cases, specifically because each of the areas has unique biophysical and ecological conditions and gradients that can be characterized and discussed in detail. Although such specific areas may not coincide exactly with the particular focus areas in which you conduct your work, by exploring and understanding the remote sensing and landscape ecology approaches in each of the case study project areas, consideration of the circumstances of each will provide you with insight into most of the relevant issues that stakeholders, managers, and researchers encounter at multiple scales, in most cases. Both spatial and temporal gradients are included and analyzed in the four examples to better characterize the environmental conditions that exist; provide a look at how complex both spatial and temporal components and scale can be influential in determining project design and outcomes; and determine how the challenges of setting these limits can be informative in your project's design. The key to understanding how the four case studies in this book integrate is to understand how scale and diversity of landscape elements, in addition to the constraints of the remote sensing data and technology, all weave together to address the key questions of decision makers.

The first of the four geographic locations selected is the Laurentian Great Lakes Watershed. This project case study demonstrates the role of landscape data in spatial and temporal ecosystems and general biophysical characterization of a very large and complex landscape. By identifying strategies for the assessments of the extent, composition, and landscape configuration of both upland and wetland elements across this vast area at a synoptic scale, the value of utilizing remote sensing to affect the evaluation of landscape ecological attributes is evaluated and demonstrated actively. The landscape metrics used are quantifiable measurements based on data that are spatially explicit and geographically referenced. In the context of watershed management, this case study can provide options for using landscape metrics, particularly in a dynamic societal context. Additionally, this study shows how relatively broad-scale use of metrics can be utilized for hypothesis generation, where remote sensing data can then be linked to field data for both validation and scaling. Ultimately, the integration of landscape scale metrics with ground-level ecological functions is critical to managing natural resources, and the work in the Great Lakes points us in that direction, which is then carried forward in later case studies.

The second selected geographic location is the California and North Carolina coastal regions of the United States. This project case study tackles a very pressing global environmental crisis, sea level rise along coastal regions of all marine coastlines of the planet, by utilizing remote sensing data and modeling of sea level data to characterize the influence on coastal landscapes, specifically low-lying areas that are typically marshy or swampy. The extent and the degree of the loss of these coastal regions is a very challenging question for geographers, ecologists, engineers, and policy makers to address in the coming years. This study demonstrates the necessity for understanding and utilizing the trade-offs available to us all in terms of the precision of various models and landscape metrics, and the need for ecologically and societally significant information to meet pressing project goals. These concepts and challenges apply equally to many similar landscape ecological issues, such as changing temperature along elevational and latitudinal gradients, and changing precipitation along elevational and latitudinal gradients, which are both also driven to some degree by climate change.

The third of the four selected geographic locations is the Ozark Mountains, United States. This project case study is a truly groundbreaking watershed analysis that redeems the goals of remote sensing being used to characterize not only the physical nature of watersheds, but also the biophysical characteristics of the contributing areas outside of the channel, riparian areas, and also by utilizing these techniques to infer the water chemistry and water quality characteristics within the water flowing through these systems. Because the analyses are moderate in geographic extent, the impression may be left that this case study is not applicable to larger areas of the landscape, but upon consideration of the methods and approaches used, it can be seen how the methods and approaches presented can be applied to any number of areas, given similar data, across a vaster area if necessary. An important step forward in this case study, as compared to the Great Lakes work, is the use of specific ground-based field ecological data to calibrate the remote-sensing based models, from the headwaters of the White River, through a multitude of tributary streams, to the main stem of the White River, which ultimately feeds the Mississippi River and Gulf of Mexico. This work also incorporates the complexity of human population increases in the region, land cover changes, and utilizes spatiotemporal complexity to inform the use of remote sensing to model ecological conditions on the ground. As in other areas that you may be considering doing similar work, the selection of landscape metrics is key to answering the necessary and pressing issues of watershed and riparian configuration of agricultural, urbanization, changes in forestland cover, and the concomitant influences on surface water conditions.

The fourth and final geographic location selected is the Missouri River and Mississippi River Watersheds, with a special focus on the Kansas River (one of twenty-two tributaries analyzed). This project case study predicts major natural hazard impacts and floods, which affect the largest number of people worldwide, averaging 99 million people per year. This area has experienced unprecedented flooding in recent decades that have caused many fatalities, evacuations, and large financial losses. In addition, urbanization in these areas is on the increase, just as in similar areas around the world. Such urbanization generally increases the size and frequency of floods and may expose communities to increasing flood hazards that result in an increasing focus by planners and land managers on the role of urbanization in the prediction of flood levels and damage. Most of this work by planners is for disaster management, as well as urban and regional planning. This representative area of flood conditions and increasing population growth and urban development in the American Midwest has faced two significant flood events in 1951 and 1993. An effective approach for utilizing remote sensing data and landscape ecological techniques to assess flood risks for people and their property can be achieved with flood risk models, which show areas prone to flooding events of known return periods. Because wetlands have the capability of short-term surface water storage, and can reduce downstream flood peaks, the use of wetland characterization and metrics can be beneficial to these ends, which is thoroughly explored at multiple scales.

Notice, as you complete the journey through Chapters 3, 4, and 5, the multitude of scales and geographic extents of the study areas and outputs, as well as the specificity of the information imparted in each of the areas. This trade-off of scale and information is a fundamental tension that exists for all work in the physical sciences, often driven by information management and the technology that allows humans to grapple with the resulting complexity. We have utilized several methods to leverage the amount of data in the largest areas (e.g., summary metrics among the Great Lakes Watersheds) so that useful ecological characterization of the landscape is provided to users, and where more information is made available by remote sensing (e.g., terrain, land cover, and land use) and field measurements (e.g., water quality, tides, and river flow), a number of statistical and geospatial modeling approaches allow for improved methods to infer ecological condition across vaster areas of the landscape.

1.6 Summary of Chapters

This book is subdivided into six chapters, each with a specific intent and fulfilling a specific piece of the larger story outlined in the previous sections. Chapter 1 is intended to prepare you for the journey through several examples of how to utilize remote sensing for landscape ecology, and set the stage for a complex topic of science, technology, and application, all within the larger context of the user's needs and the practicalities of decision making in the twenty-first century. Following the introductory work in Chapter 1, Chapter 2 provides an update of the fundamentals of remote sensing for landscape ecology, and touches on some of the relevant remote sensing advances and applications since the first edition, particularly those advances that move the science forward and have the potential for transforming our thinking in a positive way; the specifics of key technologies and data, as well as a practical concise discussion of the benefits and challenges, is included, particularly as it applies to subsequent applied demonstrations in the remainder of the book. Although many lists of remote sensing platforms and data can be compiled, the importance of Chapter 2 is the integration of the technology and data with the subtleties of project planning and the prospective focus needed in any project, with a discussion of analysis considerations that one needs to make when discovering the many platforms and data types that exist. An important topic threaded throughout the book, beginning here, is adapting to any uncertainty about platforms, data, or analytical approaches available; the question of "which are optimal for the necessary applications?" and a focus on desired outcomes are key topics dealt with in some detail in Chapter 2.

Overall, Chapter 2 should leave you as the reader with a solid basis of how you might select from the variety of remote sensing technologies and data available, from both the perspective of a decision maker and technical remote sensing practitioner or scientist; utilizing these perspectives to make excellent landscape ecology professional decisions should provide a solid basis for deciding the needs of the project prior to embarking in depth into a remote sensing–based landscape ecology project or program. Chapter 2 also presents an overview of the key landscape ecology metrics that are demonstrated in subsequent chapters. Another aspect of Chapter 2 is the general theoretical and practical pros and cons of landscape metric categories, and an outline of the benefits and challenges of utilizing selected landscape metrics, including recommendations for specific landscapes and desired outcomes, which sets the stage for the examples provided in further chapters. The latter portion of Chapter 2 provides information about the full utilization and meaningful application of remote sensing for landscape ecology and the specific considerations one must make when selecting a suite of remote sensing data types, taking into account particular ecological and biophysical environmental conditions of a landscape, an approach that allows for the inclusion of the societal factors that are always part of any landscape. This latter consideration, the societal factors involved in decision making and communication of science results, along with the technical elements of remote sensing and the discipline of landscape ecology, provide you as the reader with all of the tools needed to synthesize the several factors involved in decision making at a landscape scale. By including the practical (e.g., processing considerations) to the theoretical (e.g., scale considerations, and both spatial and temporal factors) in this manner, in Chapter 2, the comprehensive look at the options for metrics and their selection provide a solid basis for further understanding of what the best approach is for a variety of user needs, as well as an entrée to the subsequent chapters.

Chapter 3 takes a deeper dive into practical applications and provides a process study of how remote sensing–based landscape metrics might be evaluated for use, utilizing a specific example of the collaborative work that preceded the Great Lakes Restoration Initiative of 2010. Chapter 3 moves us to a thorough levels of inquiry and project development insight by providing a rich and detailed description of utilizing selected platforms, data, and methodologies to address critical and current landscape ecology research questions in the Laurentian Great Lakes, and the process evaluation has tremendous applicability to developing a collaborative use of remote sensing and other data to tackle ecological issues at multiple scales across political boundaries, which is followed up on in subsequent applied work in Chapter 4.

Chapter 4 takes an even deeper dive into practical applications and provides a process study of how remote sensing–based landscape metrics might be evaluated for use, utilizing the specific example of the collaborative work that preceded the Great Lakes Restoration Initiative of 2010. Chapter 4 moves us to a thorough level of inquiry and project development insight by providing a rich and detailed description of utilizing selected platforms, data, and methodologies to address critical and current landscape ecology research questions in the Laurentian Great Lakes; the process evaluation has tremendous applicability to developing a collaborative use of remote sensing and other data to tackle ecological issues at multiple scales across political boundaries, which is followed up on in subsequent applied work in Chapter 5. Chapter 4 also delves into another pressing issue that remote sensing and landscape ecology can serve well, that is, the influence of sea level rise on landscape futures in both the California coastal region and the coastal regions of North Carolina.

Chapter 5 focuses on regional scale landscape gradients, with the complexity of multiple watersheds and processes, through two additional case studies: the first of which is an application of multiple platforms and data sets used to inform landscape processes in the Ozark Mountains, and the second case study focuses on developing remote sensing–based landscape metrics to determine riverine (i.e., instream), riparian (i.e., river-adjacent), and floodplain conditions, impacts, and processes of the Missouri River and Mississippi River Watersheds. Both of these case studies focus on the deepest level of complexity one might find in a remote sensing–based landscape ecology project, including methods for linking remote sensing data to geostatistical inference of chemical, biological, and physical conditions on the ground and verification of those inferences, all whilst outlining the technical and logistical considerations of the remote sensing and other data approaches/needs, as in Chapter 4. The case studies in Chapter 5 are similarly specific to the others in Chapter 4, and tie in nicely as conceptual adjuncts. However, they were selected because of the tremendous need for studies and outcomes related to this type of riverine, riparian, and floodplain work throughout the world and cover some additional unique topics in those regards. Accordingly, the case studies in Chapter 5 are very translatable to the work that is likely occurring in nearly every municipality, county, state, and region of the United States currently, and are increasingly a pressing need in worldwide projects, especially the work that is desperately needed to ensure water quality and water quantity is maintained worldwide, so that water resource needs of society are met. Although regional in geographic scope, the variety of landscapes characterized, the environmental pressures, and the needs of communities considered in Chapter 5 likely apply directly to any application or circumstance that you, the reader, are working on in your project areas or may be planning to analyze in the future.

Chapter 6 takes us to the future, which is bright for remote sensing and landscape ecology, both by summarizing key themes that thread through the various case studies in Chapters 4 and 5, along with a sober look at the reality of the constraints of project development, in such a way as to catalyze ideas and creativity in the reader's application of the information and planning for their project concerns. Recommendations as to the most useful and practical applications of landscape metrics for specific future needs in landscape ecological analyses are proposed, and future approaches are matched with the potential success for developing new metrics, with a hopeful eye toward new remote sensing platforms, data, GIS approaches, and the paradigms of the future, including an analysis of the potential for integration of the concepts and practice of ecosystem services into landscape ecological projects, leveraging the rich data sets that have emanated from the remote sensing science community.

2

Key Processes for Effective Remote Sensing–Based Landscape Ecology Projects

This chapter is specifically designed to both update the detailed information provided in the first edition, and provide a solid basis for those who are seeking contemporaneous information about how to use remote sensing data for meeting the very specialized goals of present-day landscape ecology projects. In addition, this chapter outlines the current situation of the availability, utility, and potential applications of remote sensing technologies for accomplishing the diversity of landscape ecological studies that present themselves now and in the coming decades. This chapter is not intended to be fully comprehensive of all specific project constraints and goals, but utilizes a full diversity of examples and information that provides a solid outline of the information and approaches needed to increase success in designing and implementing contemporary projects with an eye toward the match between technological parameters of remote sensing and the theory and science of landscape ecology.

2.1 The Essentials

Remote sensing technologies include a variety of tools ranging in complexity from advanced sensors to simple camera and film systems. Each imaging system is as valid and important as another, depending on the usefulness for the application, and the need for creating two- to multiple-dimensional image products that can be analyzed or interpreted. The process of analyzing these images is called image processing and interpretation, which we will cover in detail, specifically as it applies to landscape ecological applications.

2.1.1 Terminology

The generally accepted definition of image or photo interpretation incorporates several steps or elements, including the fact that interpretation is both an art and science, utilized to obtain or interpret data from the characteristics of features recorded on photographs or other types of digital imagery. Photo interpretation typically refers specifically to interpretation of photographic film-based products, while image interpretation is more general, covering the interpretation of all image products including mainly digital, but also photographic, products. These terms refer to interpreting images or photographs taken from either manned or unmanned airborne or spaceborne platforms. The methods used for image interpretation are identical to the methods and skills used to obtain information from any type of photograph or digital image, and are dependent upon the interpreter's experience, general observation skills, and in the case of both human and automated interpretation, require some degree of calibration.

Image interpretation is a very important skill to have associated with your project, especially as it applies to the ecosystem types that are the focus of the work, through skillful application to obtain the necessary and relevant details about features found in images. These elements of interpretation supply information on features that are basically independent assessments of each characteristic. These elements include characteristics such as the use of color or gray tones, and the interpretation of shape, size, texture, pattern, shadow, and associations (Figure 2.1), initially, and are then informed by the materials and features on the ground. The gray tones and tonal gradations supply details about materials or features on the ground that are in a range of contrast between black and white. Typical materials exhibit a range of grey levels on black and white photographs. Rock, bare soil, concrete, and similar impervious ground materials often appear light in tone. Vegetation is relatively dark in tone due to the low relative reflectance of green plant material, compared to bare soil or rock.

The variability of visible light can be measured through the use of both monochrome (or black and white) and color film or digital imagery, to elucidate additional information about the materials and features on the ground. Color imagery shows variation of each additive primary color, such as additive ranges of blue, green and red. Color variation, like tone, can be used to evaluate materials and features as well as the spectral characteristics and variability of materials in the visible portion of the spectrum.

The shape of features in remote sensing imagery refers to the exterior configuration of objects or features. Many objects or features have distinct shapes, whether regular or irregular. Shape can be a unique clue as to the identity of the feature. Again, regardless of whether the analyst is utilizing film-based remote sensing or digital imagery, this and the other interpretive elements are important, both for human interpretation and automated processing and pre- or post-processing analyses. Size refers to the absolute or relative dimensions of the object or feature (Figure 2.2). Size can be very important as features are often indistinct on photographs and knowledge of size can be a big clue as to the identity of a given feature or the potential relationship with human activities. Human activities are typically, but not always, the source of regular shapes and patterns observed in remote sensing imagery.

FIGURE 2.1
The use of color or gray tones, and the interpretation of shape, size, texture, pattern, shadow, and associations is a fundamental element of remote sensing, and expertise in field ecology helps in interpreting imagery for landscape patterns and processes.

Texture is a variation in tone or color caused by a mixture of materials at a given location. Image interpreters or automated algorithms usually identify a texture as a subtle to marked variation in tone or color. For example, in an old abandoned farm field the texture may result from a tonal or color variation that resulted from the presence of shrubs or small trees in the field. In the eastern United States, such trees may be sumac or aspen, and in the western or mountainous United States, they may be small conifer trees or shrubs. These shrubs or trees are too small to recognize by their individual crown or canopy vegetation, but the variability in texture provides context

FIGURE 2.2
The size of elements within imagery is an important clue for interpreting human uses within a large landscape.

enough to the image interpreter to annotate the area as mixed shrubs and/or trees. The mixture of vegetation contributes to the overall variability of tone (Figure 2.1), as compared to the relatively uniform tone of a mature soybean or wheat crop in an adjacent field, for example. The presence of tall plants, partial canopy closure, and shadows helps to create a textural difference that often allows for the visual separation of natural vegetation from crops, and indeed in distinguishing between corn crops and soybean crops.

Pattern is also a very important measure, which can be intuitive to recognize for some image interpreters, developed through visual training and experience for others, or enhanced and developed through a number of auto-mated means (e.g., eCognition software, http://www.gim.be/en/products-services/gis-software/ecognition, checked January 25, 2017). Pattern refers to the regular or irregular distribution of features on the Earth's surface. An example is the regular pattern of agricultural field boundaries in a rural area. When this pattern is disturbed, for example by a stream course, one can infer certain characteristics of the stream channel—such that it is too big to be altered by human activities in the past and over time (Figure 2.3). Often these channels will be obscured by vegetation on the banks of the channel, in the riparian vegetated zone, such that the water itself is difficult to see. The pattern of riparian vegetation can be used to generally track the stream course because it indicates the linear curvature of the stream feature. In com-bination, the stream feature and the vegetated riparian areas may indicate the presence of shaded stream areas that offer key habitat for stream biota.

The shadow of a feature can also be an important clue in determining what it is, and shadows may help identify the feature regarding its material(s) too. Shadows cast by a feature can provide important shape and size informa-tion. A simple example is that of human interpretation of the name of a busi-ness or building from the shadows of individual letters that form a sign. The individual letters blend into the building, but the shadows are distinct to the interpreter or may be too narrow to view from a remote sensing platform,

FIGURE 2.3
Patterns within the landscape are telltale signs of not just the identity of a landscape element, but also the ecological or other processes within a large landscape.

such as an aircraft or satellite. Conversely, shadows can obscure detail. It is difficult to view a feature hidden by a shadow due to the lower relative illumination in the shadow area as compared with the overall illumination of the sunlit areas.

The association of a feature is the other characteristic or clue that is found together, in association, with the feature of interest. For example, the course of a stream is usually evident by shape, but often the tone of the water is hidden from view by trees. Stream courses are usually found in association with other clues, such as the meandering stream pattern; a branching drainage pattern; automobile and railroad bridges; ponds or lakes; stream-side vegetation; and lower relative elevation or a downhill course, as compared to the surrounding landscape (Figure 2.2). All of these fundamental elements assist in the overall interpretation and ultimately the quality of measurements and interpretations across a landscape, which, when accumulated across vast areas such as are described in specific examples in Chapters 4 and 5, must be sound and accurate.

2.1.2 Image Interpretation

Interpretation of imagery can supply high-quality detail for water features, wetlands, forests, urban/built areas, and other land cover or land use types. When you take full advantage of image interpretation, along with specific accumulated knowledge of the features of interest and the experience necessary to make accurate interpretations, you can save tremendous amounts of time in characterizing landscapes and features of interest, which would otherwise require ground crew time and other efforts that are extremely costly for most applications and geographies of interest.

When viewing of images is necessary for the accomplishment of a project, it is widely understood that skilled interpretation provides the specialized interpretation needed, and characterizing complex landscapes, such as wetlands (complex transitional ecosystems) (Lopez et al. 2013) or aquatic ecosystems (Niedzwiedz and Ganske 1991) is optimized by such specialized skills. Almost all landscapes have a need for this specialized set of skills due to the ubiquity of wetlands and aquatic ecosystems worldwide. One can see details of the nearshore topography or bathymetry as a result of the penetration of visible light into, and reflected back from, the water. Upland areas of landscapes are typically less difficult to characterize than (transitional) wetland areas and aquatic areas, primarily because they do not have the reflectance variability of water found in wetland and aquatic ecosystems. When focusing on upland areas without wetlands or aquatic ecosystems, such as upland forests or urban areas, much of the characterization is focused, initially, upon whether vegetation is absent, grassy, shrubby (e.g., low or sparse wooded), or forested (e.g., tall or dense wooded).

Remote sensing–based ecological interpretations consist of utilizing evidence that can be accumulated and articulated for a given ecosystem or landscape under study, and can also be formalized into a full list of characteristics or interpretation key to facilitate identification and inventory. This approach of preparing an interpretation key has worked well in a number of instances, which may or may not apply to your project work, but is advised in most instances and used to good effect to characterize a variety of vegetation types, including grasslands, shrublands, forest, wetlands, or other aquatic ecosystems in a complex and vast landscape. For example, submerged aquatic vegetation is among the many focal areas of habitat in coastal areas, and these areas have benefited from this approach of focused and organized interpretive work (e.g., Raabe and Stumpf 1995). Focused work on mapping submerged aquatic vegetation has resulted in a beneficial characterization of the Chesapeake Bay in the United States during a massive ecosystem restoration initiative begun in the 1980s (Ackleson and Klemas 1987), and currently benefits the ecology of the area, and by connection the communities of this large area of the landscape have seen an improvement in these critically important aquatic habitat areas. Specifically, the use of remote sensing and landscape ecology provided new information about the differences in spectral reflectance, which importantly facilitated a number of landscape restoration efforts that benefitted the health and well-being of the ecosystems and communities of the Chesapeake Bay region, such as is related to fisheries and recreational/aesthetic aspects of the area. The technical analysis of the reflectance characteristics for the Chesapeake Bay project utilized the fact that dark-toned features found beneath or above the water, emergent and submergent vegetation, appear dark because plants reflect little light in the blue part of the visible spectrum and in the red part of the visible spectrum. Thus, in black and white or color images of these plants, they will be observed as darker tones or darker colors, respectively, as the green reflectance of light by plants is relatively low compared to that of soil. This important approach for organized interpretation of remote sensing data as it applies to ecosystem condition is one specific example of organizing an effective landscape ecology project, which can be replicated in many other circumstances.

Because interpretations of images are enabled by the active use of the differential spectral reflectance in different portions of the electromagnetic spectrum, the use of multiple images of different film types (e.g., infrared color, visible color, black and white) for the same area of the landscape can be a great help in providing depth of interpretations and details that might otherwise not be apparent (Lyon 1987, 1993; Williams and Lyon 1991; Lyon and Greene 1992a; Lyon and McCarthy 1995; Lillesand et al. 2014). This is particularly true for the characterization of hydrological elements in areas that change over short or long periods of time. Multiple dates of photos or images and multiple types of films or sensor band data can help tease out the intricacies of apparently general hydrologic characteristics, mainly by allowing the detection and distinction of wet, dry, and intermediately moist areas during different periods of time.

Following rainstorms and other hydrologic events that follow, such as flooding and erosion, areas of water may be opaque due to the runoff and suspension of sediments in the water column. These suspended sediments (where applicable, also referred to as nonpoint-source pollutants) may be particularly difficult to image through because of water column opacity, and so suspended sediment can disrupt evaluations of submergent materials and aquatic ecosystem features that would otherwise be visible in clearer water-bodies, such as vegetation growing beneath the water's surface (Lyon et al. 1988) or other materials beneath the water's surface. Conversely, suspended sediments in water may help to locate the movement of water and water-borne pollutants along a river course, or in a bay, for example. As mentioned earlier, multiple images from either different sensors or times may help to lend insight into a variety of aquatic or other landscape conditions by capturing variability in the imagery over time. In this way, utilizing the opacity of water, or lack thereof, serves several needs and the techniques of interpretation, which can lead to some creativity in terms of applications to landscape ecological questions posed, and project designs.

It is always recommended to become cognizant, and possibly obtain (depending upon the cost and need), multiple dates of aerial photos or digital imagery of a focal landscape, watershed, or study area in support of ecological analyses, when possible. This is because each image supplies unique information and multiple coverage of imagery adds the value of repeated samples. The repeated samples provide increased statistical power, as with all sampling designs, by an additive informational approach. The depth of knowledge for an entire landscape ecological project can be thus tremendously improved with the use of the informational power of multiple images. Although commercial imagery can be costly, there are inexpensive imagery options to explore and often these options are historical in nature. Obtaining these historical images, particularly if there are multiple dates, can provide important contextual information that can supplement other field data. Collaborative approaches to sharing the costs of such endeavors, or utilizing freely available public data, is a very wise approach, and is a major element of all of the examples utilized in Chapters 4 and 5.

Imagery that is either orthorectified or otherwise quantifiable in terms of distortion is also valuable in determining the most accurate location of the position of features and materials (O'Hara et al. 2010). Knowledge of size and position of landscape features can be supplied by specific measurements within images (Figure 2.4). Most simply, analysis of images can provide linear and areal measurement of features by utilizing both photogrammetry or surveying technologies. Photogrammetric measurements can be made on the images, and tied to some absolute reference to characterize the location of earth or terrestrial features, empirically (Velpuri et al. 2009). Field surveying also allows one to later relocate these features using the original measurements or map products (Van Sickle 2008).

FIGURE 2.4
A fundamental tool for landscape ecologists is the measurement capability of imagery, utilizing the geometry of the captured data from a remote sensing platform.

2.2 The Key Data Types

The number and type of digital imagery data are numerous, and growing, and the number and type of analogue (i.e., film based) imagery types are also variable in that they were (a few still are) typically collected in a variety of ways and contain unique and irreplaceable information. There are also secondary data products (maps and models derived from the digital and analogue imagery) that are equally important to be aware of, and skilled at utilizing to their fullest extent. Both analogue and digital data are important to consider for use, and neither are superior to the other, which is a common misperception. Typically, analogue data provides excellent historical information, where digital data may provide better precision and perhaps coverage, although these are not steadfast rules. It is recommended that all secondary products be utilized after understanding how they were derived, including data types used, processing approaches, and any assumptions used during their production. The remainder of the book is replete with examples of existing and potential sources of both airborne- and satellite-based remote sensing data, including sources. Selected examples are listed in a non-comprehensive listing below:

- Alaska Satellite Facility (http://www.asf.alaska.edu)
- G-LiHT (http://gliht.gsfc.nasa.gov)
- GLOVIS (http://glovis.usgs.gov/)
- LVIS (http://lvis.gsfc.nasa.gov)
- MassGIS (https://wiki.state.ma.us/confluence/display/massgis/Home)
- NASA (https://earthdata.nasa.gov/user-resources/remote-sensors)
- NASA EOSDIS (http://earthdata.nasa.gov)
- NEON (http://www.neonscience.org/science-design/collection-methods/airborne-remotesensing)
- Oak Ridge National Laboratory DAAC for Biogeochemical Dynamics (http://daac.ornl.gov)
- Sociological Data (http://sedac.ciesin.columbia.edu/theme/remote-sensing/data/sets/browse)
- USGS Earthexplorer (http://earthexplorer.usgs.gov)

2.2.1 Aerial Photographs

Most data collection in the twenty-first century has consisted of digital data, although some applications of (simultaneously collected) film aerial photography does exist in the twenty-first century and some specialized collections are still being made. Although often ignored, likely due to the popularity and attractiveness of contemporaneous digital data sources (especially download-able and online versions of these digital data), this extremely valuable resource of (now, typically historical) aerial photography is necessary to better understand the context of the contemporary landscape conditions, especially for any information that precedes the digital era. Aerial photographic coverage of the United States is available from archives, and are often now available online (e.g., Google Earth Professional, https://www.google.com/earth/download/gep/agree.html, checked January 25, 2017). Generally, historical photographs in the United States date back to the late 1930s, and in some cases, U.S. Army aerial photography dates back to 1916 in specific locations. Multiple sets of aerial photos are in data archives, and one may be able to develop a time series of photos extending over the decades until the present with overlays, which can be aided by digitization (e.g., image capture by a high-resolution flatbed scanner) and software that "warps" or "rubber sheets" several different oblique aerial angles together for easier interpretation, and possibly by measurement or other quantitative comparisons; these are very specialized approaches that may have application for viewing, processing, and integration with more contemporary digital imagery for a particular landscape.

The interpretation of historical aerial photos provides data on a number of conditions. From individual dates of coverage, interpretation will yield data on the land cover types, presence or absence of houses and buildings, stream drainage pattern, and general soils and geomorphology characteristics. For example, multiple dates of coverage allow the user to capture the different hydrologic conditions of forested areas, wetlands, lakes, urban and other built areas, and coastal zones that have occurred and possibly changed over time (Butera 1983; Lyon et al. 1986; USACE 1987; Carter 1990; USEPA 1991; Williams and Lyon 1991; Lyon and Greene 1992b; Lyon 1993).

Sources of historical aerial photographs at the federal level include the National Archives and Records Service in Washington, DC, for pre–World War II photographs; the U.S. Geological Survey EROS Data Center in Sioux Falls, South Dakota, for U.S. Department of Interior agency photographs; and the Aerial Photography Field Office in Salt Lake City, Utah, for U.S. Department of Agriculture (USDA) agency photographs. The addresses for these sources are publicly available at each agency's Internet site and in Lyon (1993), Lyon and McCarthy (1995), Ward and Trimble (2003), and from other sources outlined in this book. Aerial photographs or digital imagery are usually available for every other year or every third year in the U.S. Geological Survey (USGS) and USDA archives. It is possible to gather approximately ten or more dates of aerial coverages since the 1930s, consistently across the United States, from the sources above (Lyon 1981; Lyon and Drobney 1984).

An excellent source of aerial photography is the U.S. Department of Agriculture's Farm Service Agency. Since approximately 1981, the agency collected small format (35-mm) color transparency aerial photographs of farmed areas subject to crop support programs (Lyon et al. 1986). Currently, these data are included in the USDA aerial imagery portal that combines all of the USDA agency programs (https://www.fsa.usda.gov/programs-and-services /aerial-photography/imagery-programs/naip-imagery/, checked January 25, 2017). The photos from the USDA resource portal include those that were taken early on film and acquired at relatively low altitude creating large-scale photos for analysis, and now comprise an excellent archive for landscape ecological analyses. Significantly, these images have been acquired during the growing season of crops and other vegetation, as well as regularly during the dormant or leaf-off period of vegetation. Leaf-off aerial photographs are valuable in that land surface details can be observed and not obscured by leafy vegetation (USACE 1993). It is important to not cast aside these resources because they have tremendous application to landscape ecological efforts. For example, it is possible to identify and inventory wetland areas at a fine scale using leaf-off season photographs, allowing for a fine resolution look at the extent of wetland areas, or areas of past wetland locations within farm field, often definable when fields are fallow and there is minimal vegetation residue in crop fields. Thus, from the differences observed in the imagery between these residue areas and the wetlands or other adjacent terrestrial or aquatic ecosystems, a tremendous amount of ecological information can be ascertained, with some effort (Lyon 1993; Lyon and Lyon 2011).

2.2.2 Specialized Digital Imagery

The continued introduction of advanced sensors for landscape ecology has made the issues of identifying and characterizing the variability within and among ecosystems more feasible (Antolovich 2011; Honkavaara et al. 2013). Among the reasons for this increase in feasibility is that the simultaneous use of multiple parts of the electromagnetic spectrum and the availability of finer spatial resolution data has enhanced these capabilities immensely (Gilmore et al. 2009; Klemas 2011; Nagendraa et al. 2013).

There are still the challenges of seasonal variability and limited coverage, as well as limited availability, of historical data that hampers the capability to conduct perfect change analyses, as well as spatial resolution limitations. Some advances to address these issues have occurred with frequent coverages by certain sensors, such as the Moderate Resolution Imaging Spectroradiometer (MODIS; https://modis.gsfc.nasa.gov/data/; checked January 25, 2017) to follow the short-frequency variability of vegetational and hydrologic change (Thenkabail et al. 2005; Linderman et al. 2010; Lunetta et al. 2010; Sun et al. 2012).

The capability of spaceborne sensors of high spectral fidelity and fine spatial resolution are important to take advantage of in landscape ecological applications. There are a number of key examples of these sensors, including MODIS, and the Advanced Spaceborne Thermal Emission and Reflection radiometer (ASTER; Thenkabail et al. 2005; Wolter et al. 2005; Callan and Mark 2008; Pantaleoni et al. 2009; Yang et al. 2009; Akins et al. 2010; Rodrigues-Galiano et al. 2012). Hyperspectral sensors (typically considered to be sensors that collect in the tens to hundreds of narrow spectral bands nearly simultaneously) are also demonstrating great promise for identifying the composition of vegetation on other ecosystem environmental conditions (Lunetta et al. 2009; Zomer et al. 2009; Thenkabail et al. 2012; Hinckley et al. 2016).

The utility of Radio Detection and Ranging (RADAR), or microwave, sensors is particularly attractive because they provide their own source of microwave radiation to illuminate a target and a major advantage is the capability to penetrate through cloud cover and most weather conditions. Because RADAR is an active sensor, it can also be used to image the surface at any time, day or night (Lyon and McCarthy 1981; Wu 1989; Ramsey et al. 1999; Islam et al. 2008; Lang et al. 2008; Betbedera et al. 2015). Most RADAR wavelengths respond to the dielectric constant, or conductivity of the materials, and thus to the roughness of vegetational canopy structure, relative to the wavelength (Touzi et al. 2007; Ramsey et al. 2011a,b). As such, RADAR is an excellent adjunct to visible and infrared sensors for characterizing vegetated landscape conditions. Examples of RADAR utilized for mapping include the Canadian Space Agency's RADARSAT-1 and RADARSAT-2. The integration of other sensors, such as Light Detection and Ranging (LiDAR), has been particularly useful by adding a topographic component, as well canopy height and other structural characteristics (Ramsey and Jensen 1995; Ramsey et al. 1998, 2004; Asner 2011; Asner et al. 2008, 2011; Yang et al. 2009; Gonzaleza et al. 2010; Yang and Artigas 2010; Cho et al. 2012).

The previous concise summary of some of the more highly specialized sensors, and sources for exploring those data available for landscape ecology applications, is intended to generally present methods and approaches to enhance standard remote sensing data in order to detect and characterize all ecosystems. Regardless of the sensor type or capability, outputs from these sensors can provide input to a GIS for beginning the mapping and modeling stages of the work (for example, Yi et al. 1994; Maidment and Djokic 2000; Lopez et al. 2003; Lyon 2003; Martz and Garbrecht 2003; Mehaffey et al. 2005; Thenkabail et al. 2009; http://www.earthobservations.org/index.shtml and http://www.usgeo.gov, checked January 25, 2017). Subsequently, specialized sensor products can be assessed as to their accuracy and precision, and land cover types can be addressed through producer and user accuracy (Congalton and Green 2009; Burnicki 2011). Demonstrating the uses of remote sensing data, and coordinating those uses with field work in landscape ecology applications (for example, Van Derventer 1992; Shuman and Ambrose 2003; Ward and Trimble 2003; Stevens and Jensen 2007) is the most effective way to ensure integration into assessment tools, for optimal project success.

2.3 Meeting the Ecological and Societal Needs of a Project

Utilizing periodic monitoring with aerial photographic, aerial videographic, or aerial or satellite remote sensing digital imagers provides the necessary periodic information for cost-effective assessments, which otherwise would take tremendous, nearly impossible, amounts of labor to maintain in any reasonable area of the landscape, much less a vast region (such as are described in the case studies in Chapters 3 through 5).

As we have just reviewed in the previous sections, utilizing knowledge of the differences in tone, color (or digital representation), and other imagery characteristics can aid in the interpretation of static images, or with multiple images of the same location on the ground during different weeks, seasons, years, or decades. The difference in tone or color can be demonstrated, or quantified, from digital data. When a forest exists, the tone can appear dark on black-and-white images and on color images the addition of green to this dark area adds context for the interpreter. When houses are constructed in, among, or nearby a forested area, the tone during construction is light on black and white images and the color is brown-white on color images. On color infrared (CIR) film or sensor imagery, bare soil areas appear blue-green in color and light-toned or bright.

Although the differences in light reflected from one kind of land cover or land use as compared to another may be used as an indicator of what it is on the ground (i.e., interpreting the materials on the ground, from above) and if change has occurred (Lyon et al. 1998) the interpretations of those elements in a way that is meaningful both ecologically, and sociologically, can differ. Landscape ecological parameters and processes can be well characterized, monitored, and assessed using both optical film and optical or non-optical sensor remote sensing data, using either airborne- or satellite-based platforms; however, the manner in which those elements are combined and portrayed (i.e., communicated) to decision makers and community members is important. The remote sensing essentials outlined thus far cannot ensure that these important steps are taken; however, by utilizing imagery, modeling, and mapping in a variety of ways, we can ensure that we meet the specific ecological and sociological needs of the particular landscape, watershed, or communities involved. In this way, we can also begin to ensure that the geospatial knowledge and theory, along with the data and technology that are the foundations of remote sensing, are effectively combined to specifically address the societal needs of landscape ecology. This is the subject of subsequent sections of this book, which include some practical constraints of conducting remote sensing and landscape ecology project work in today's world.

2.3.1 Data Cost, Availability, and Quality Considerations

Decisions about which type of ecological information, remote sensing data, and GIS data to use in order to execute a successful landscape ecology project are complex, and require forethought and optimization of three important factors:

1. The cost of data (i.e., acquisition, processing, and storage)
2. The availability of the necessary data
3. The quality of the data

In the planning of a landscape ecological assessment, whether for ecological research, regulatory support, or other decision making, one has to decide, for example, between the use of an objective data source with high quality but many gaps in coverage, which would require a large portion of available resources to collect sufficient data, or the use of other data sources, with fewer gaps and less costly information, but requiring a reduction in reliability and comparability.

Decisions about how to evaluate and monitor ecological conditions at a landscape scale must always take into consideration the logistical challenges presented by larger areal extents and the fact that landscape assessment and monitoring schemes must be parsimonious. Although there are many benefits associated with exploiting existing data, there are costs (e.g., non-contemporaneous data incompatibility) that must be considered in accessing and processing those data. Although long-term or large-extent data are generally accessible through major data centers, bringing these data sets together and ensuring compatibility is always an important time sink for most projects. Short-term or single-site data sets, often generated to address focused scientific questions, short-term monitoring needs, or preliminary inquiries are often available solely from the originating organization and willingness to share for those who collected the data is limited because of their investment or proprietary needs or concerns about sharing the primary data. One potential solution in these situations is to partner with the originating organization and inquire whether they are interested in partnering on additional work that you can provide to them or inquire if they are interested in providing secondary remote sensing–based products (maps, models, or other outputs necessary for your project) at a nominal cost. Keep in mind, as with all products, the overall cost of any landscape assessment is affected by the availability of existing data, its source (whether from a public, nonprofit agency, or from private for-profit companies), and its quality. A strategy for success is to think of what you or your organization bring to the table that would be valuable from the other side's (e.g., the data producer's or owner's) perspective. This may include staff time from your organization or allied partners in terms of data processing, field-based work that is of value to both of you, or other equivalently valuable work, equipment, or materials.

Existing data are often fragmented and dispersed among many sources, depending on the geographic and environmental areas that are considered. This issue can be especially relevant when local information is needed at a broad scale for land management, for example in the Great Lakes Basin assessments (Chapters 3 and 4). In addition, supporting databases for landscape ecology analyses, often from the same data type, are created in several formats or geographic projections that may not be interoperable, such as was the case in the Ozark Mountains field-based water quality data and imagery (Chapter 5). These availability and interoperability issues are all very important issues to understand prior to the selection of the data or metrics for a project.

U.S. federal agencies are likely to be the primary lower cost sources for data that include maps of elevation, watershed boundaries, road and river locations, human population, soils, land cover, air pollution, and more parameters. Some typical governmental sources in the United States include the U.S. Army Corps of Engineers, National Oceanographic and Atmospheric Administration, U.S. Geological Survey, U.S. Environmental Protection Agency, U.S. Department of Agriculture, U.S. Census Bureau, and the Multi-Resolution Land Characteristics Consortium (multiple agencies). Resources available from these sources typically consist of databases, raw or preprocessed remote sensing data, digitized maps, and GIS/statistical models or software. Data types that use standard methodologies, such as those that comply with the U.S. standards for these data types, and thus the standards of the Federal Geographic Data Committee (FGDC), provide data for which reliability is relatively high (or as least well documented), availability is assured, and long-term updates and other data maintenance is assured (Federal Geographic Data Committee 2010). In the absence of these data quality assurances, it is not recommended that long-term landscape ecological analyses be initiated since there may be future interruptions in data, nor is it prudent to utilize non-FGDC data for any landscape ecology project that may come under regulatory, legal, or liability scrutiny in the future.

2.3.2 Selecting among Key Data Types

The types of data that may be available for landscape indicator development include ecological information, remote sensing data, and other geospatial information. Ecologists have traditionally used historical maps, aerial photographs, and their understanding of spatial relationships between ecosystem patches to explore relatively broad-scale ecological characteristics of the landscape (e.g., Miller and Egler 1950; MacArthur and Wilson 1967; Howard 1970). These techniques are long standing and have stood the test of time since their original inception.

The inherent complexity of ecosystems, and the many specific ecological processes that are contained within them, have prompted development of, and research into, the use of specialized airborne and satellite sensors, and related processing techniques for those new data types, as they have come online and become more and more available to us all in the twenty-first century. Within the past decade environmental scientists have successfully integrated and applied the use of relatively sophisticated sensors (e.g., both airborne and spaceborne multispectral and hyperspectral; e.g., Johansen et al. 2010), automated image processing software/techniques (e.g., ERDAS Imagine, http://www.hexagongeospatial.com/products/power-portfolio/erdas-imagine; ENVI software, http://www.harrisgeospatial.com/ProductsandSolutions/GeospatialProducts/ENVI.aspx, checked January 25, 2017), and the computing power of GIS (e.g., ESRI's ArcGIS, http://www.esri.com, checked January 25, 2017) to the study of ecology.

Because of the variety of data types and their availability, which has skyrocketed in the past 20 years, and the software that can be procured to process the data, which has undergone tremendous advances in hardware and software in the past 20 years, landscape ecology project concepts are now often developed along with the public and often envisioned or requested specifically by community members or decision makers. Frequently, less trained individuals may gain access to online remote sensing data and begin to purchase, download, and view data, regardless of adequate experience. At times, one or more members of a team who may have some exposure to GIS may begin initial data manipulation and analyses. Naturally, the skills required to process and interpret remote sensing data do not always reside with GIS professionals because of the difference in training and skill sets needed for each of these two very important and specific disciplines. Similarly, many landscape ecologists have a background in understanding the use of geospatial data, yet do not have the specific current training for GIS and remote sensing data processing. For these reasons, it is important that the full team of landscape ecologist, GIS specialist, remote sensing scientist, other key scientists or statisticians, and, importantly, the customer or decision-making body/community be equally involved in the selection of data or be involved in the outcomes of the selection decisions. In this manner, the transdisciplinary (Hirsch Hadorn et al. 2006) nature of landscape ecology can be retained, and include the technology and scientific expertise needed to address all of the elements of any landscape scale project.

2.3.3 In Pursuit of Ecological Information

Sampling ecosystem components, such as water, soil, air, plants, and organisms on the ground is a very traditional approach to determining the ecological functions in a particular area; a very well-established approach. However, ecosystems are impossible to survey, that is, to have a complete measure of the area, which is why sampling has been the necessary approach to estimate what might be found as a full measurement of all of these components throughout the entire ecosystem or landscape. Thus, an ecological study must always consider what components of an ecosystem to assess in order to answer the fundamental question that is at the heart of broader-scale projects; that is, "Which ecological indicators are the best surrogates of overall ecosystem condition?" A traditional ecological approach might be to consider a good ecosystem indicator as one that can be explained in its component parts, which is known a priori, to be directly linked to the functional status of the ecosystem (e.g., hydrologic condition), and has demonstrable and repeatable linkages with the functional status of the ecosystem. A landscape ecologist might additionally consider a good ecological indicator as one that reflects conditions across multiple ecosystems, multiple watersheds, or at other broader scales. Environmental policy experts may be additionally interested in selecting ecological indicators that have the greatest likelihood of answering the following related questions, which they and society typically need to address:

- What indicators characterize and measure ecological sustainability?
- What indicators best show changes related to human activities?
- How can indicators developed in one place and time be used in other places and times?

To some extent, different measures and monitoring designs are needed to answer all of these questions, and to answer them at the relevant local, watershed, regional, national, or global scale(s). While local or watershed assessments may include fairly complete monitoring of landscape stressors and impacts, such direct assessment is clearly not practical over large regions, using the same rationale for sampling, rather than surveying. However, utilizing a number of mathematical and geostatistical methods outlined in Chapters 4 and 5, there are opportunities to harmonize assessments across large landscape extents by including information at the field-based scale to approximate and predict full wall-to-wall landscape conditions. Although this approach is not a true survey of the landscape, it does provide a probability of conditions for every reporting unit of an assessed landscape area.

2.4 Metric Measurability, Applicability, and Sensitivity

After reviewing data availability, types, and the associated considerations of their usage for a specified project, a next step is the selection of landscape metrics, which requires considering three important questions:

1. Are the available data capable of adequately measuring the (metric) parameters, and do they address the ecological endpoints of interest from an ecological standpoint?

2. Are the metrics to be derived during the landscape ecological analyses applicable to the ecological endpoints of interest, and do these results answer the questions of the audience (e.g., policy makers, community members, scientists) for the analyses?

3. Are the metrics to be derived during the course of the landscape ecological analyses likely to be sensitive enough to provide information about the ecological endpoint(s) of interest?

Evaluation of measurability of a landscape metric must include a primary review of the expertise, training, and methodologies used to acquire and process the remote sensing data, input and analyze the derived GIS data, and synthesize the results of such analyses. These three measurability-related steps typically require the input from individuals that have expertise in remote sensing, computer science, geography, GIS, landscape ecology, general ecology, environmental science, chemistry, hydrology, geology, and other relevant specialized fields. Evaluating measurability also involves an early review of prior techniques, and determining how they may be modified to accomplish the particular goals of a landscape scale ecological project. It is often tempting to repeat some of the same techniques applied in other geographic locations or in other ecosystem types, but it is important to always parameter-ize (i.e., calibrate) your work to the specific biophysical conditions and goals of the landscape under consideration, using the generalized approaches described above.

The applicability of a landscape metric (or indicator, derived from that met-ric) is also a critical step that a research team should address prior to begin-ning the landscape assessment process. For example, in the Great Lakes, the State of the Lakes Ecosystem Conference (SOLEC) has compiled several lists of operational and proposed field-based and other measurements that are applicable to landscape assessments, which can be used to develop landscape indicators (see detailed discussion in Chapter 3). Not all of these measure-ments have been completed at a broad scale and some have been completed solely at selected sites around the Great Lakes. Additional information about these measurements may be used as analogues for applicability of landscape scale metrics for the development of landscape indicators in other areas; however, calibrating metrics from measurements in other geographies is necessary to validate the work elsewhere.

Landscape metrics and metric-derived indicators must demonstrate that they are sensitive to (spatial and temporal) changes that occur specifically in ecosystems of a particular area of the world if they are to provide infor-mation to the relevant conditions and societal needs, determined in the measurability evaluation. Because all ecological systems change, metric and indicator sensitivity must be gauged at a relevant spatial and temporal scale that makes sense for the ecological endpoint (Bailey 2009), determined in the applicability evaluation. Metric sensitivity can be evaluated by field veri-fication and then further evaluated (as a landscape indicator) by validating a response by the metric to known (and validated) drivers within, or in the vicinity of focal ecosystems. For example, a landscape metric that approxi-mates nutrient conditions in wetlands of two different watersheds can be field-tested using a statistically valid field sample of wetland soil and water chemistry in those two watersheds, and then could further be tested as an indicator by field validating the relative proximity of agricultural land to those wetlands throughout each of the watersheds.

2.5 Associations between Remote Sensing and Ecological Data

As discussed previously, because of the vast areas involved, and the com-plexity of information that is required to assess the ecological functions of large areas, remote sensing technologies have provided new opportunities and additional sources of information to develop indicators across broad areas of the landscape. However, for the reasons also outlined previously, one's ability to interpret landscape spatial patterns and identify materials on the ground can be a challenge because of the limits of the spectral (e.g., detectable energy wavelengths) and spatial (e.g., minimum pixel size) properties of any particular sensor type. Perhaps for these reasons, remote sensing data are mistakenly thought to be less useful than more proximal observations made on the ground within the ecosystems of interest. However, field-based measurements (e.g., 1,000 sample points within a forest) may not be as comprehensive for an entire area of the landscape as remote sensing data and thus may be relatively less effective at determining the true type, number, distribution, and functions of some of the key elements of an ecosystem (e.g., plant species distribution or degree of fragmentation of a large forested area of the landscape). Consequently, remote sensing data can supplement the inability of investigators to effectively sample a wetland by providing important wall-to-wall coverage that can be utilized to fully assess multiple ecosystems across a vast landscape. The inherent trade-off between having a full coverage of remote sensing data with likely less detail, and having a more proximal and detailed set of field-based samples, is not solvable and explains a basic tenet of landscape ecology, which is that there is no particular value for data of one scale that can be directly compared with the value of data at another scale. This multiscale trade-off as it relates to data is the reason why both remote sensing data (or derived geospatial data) and field-based ecological information are key elements needed to solve questions among scales, and why all usable and valid data at multiple scales should be thought of as complementary to each other. It is therefore very important to carefully and thoughtfully select the necessary data for a particular set of landscape ecological questions, which may dictate what scale and specificity of data is most useful for that application.

Because it is so important to initially determine the scale of ecological information that is necessary to assess ecosystems at a landscape scale, prior examination of needs and a considered determination of which types of remote sensing data are required for an effective and successful assessment project. For example, if the goal is to measure simple spectral or spatial characteristics (e.g., measuring for the presence or absence of large forested or large open water areas) then it is not necessary to have fine spectral or spatial resolution information, and relatively coarse resolution satellite data (e.g., MODIS or Landsat) could suffice. If the goal is to measure more complex characteristics of ecosystems (e.g., the presence and area of herbaceous vegetation in lakes), then it may be necessary to acquire data with higher spectral and spatial resolution for those areas of vegetation, provided the data can still be used to cost-effectively cover a vast area, if needed. The appropriate spatial resolution of remote sensing data is thus determined by the investigator deciding what the minimum conceivable level of spatial information is necessary to adequately assess ecosystems and then determining the optimal pixel size that provides that information, in the context of the cost and availability factors. The spectral resolution necessary for the landscape assessment is also determined through a similarly thoughtful and reasoned approach by the investigator or team who is deciding what the minimum conceivable level of spectral information (e.g., to determine water stress in plants through the reflectance values of plant leaves) is necessary to adequately assess the condition of an ecosystem and its elements. The ecological investigator or team should also determine whether there is a need for temporal data analyses in the future, and at what frequency the remote sensing acquisition might be required to adequately assess the condition of an ecosystem. All remote sensing platforms have differing return rates (i.e., repeated overflight), ranging from approximately daily to once every several weeks, and naturally these factors impact all of the cost functions for collection, processing, and interpretation.

Since the characteristics of land cover or land use in the vicinity of any ecosystem can determine the condition of the ecosystem, and these relationships are often quite important to decision makers since these conditions may change over time, it is most advantageous to assess these features repeatedly over time using remote sensing to meet project goals and to ensure results are updatable and relevant as the project timeline proceeds. Such temporal change analysis approaches are particularly important because, as human societies develop and increase built areas, stressors related to land cover and land use change are often those related to human activities and endeavors, such as agriculture, urbanization, and industrial development, which are all activities that can occur at rapid rates or are stochastic. We have a long and rich set of observations that have been recorded over the past two centuries, using ground and aerial photographic, or other remote sensing or mapping records, of human-induced disturbances in ecosystems across the world, including those associated with upland development, shoreline development, deforestation, changes in upland agricultural practices, road construction, dam construction, war, hydrologic alterations, and more. These changes or activities continue to be directly observed or inferred using remote sensing. Aerial photographs (generally 1-meter spatial resolution) or airborne digital data (generally a 1–5-meter spatial resolution) are often available for specific areas and can be used to correlate land use changes with ecosystem alterations, and for broad-scale applications the use of satellite imagery up to 30 m has been used routinely to assess these changes. We continue this tradition of measuring such changes into the future, and the tools we are using in the current era are recording these events now, nearly in real time.

2.6 Leveraging Land Cover Products

Remote sensing data are the source for much of the derived land cover and land use data sets that are used for GIS analyses and modeling. Generally, GIS products are derived by either manual remote sensing data interpretation or semi-automated image processing. The examples provided in subsequent chapters comprise the bulk of the biophysical characteristics of the landscape

under study, and all lead to the generation of landscape metrics and indicators that were developed using the wide array of spectral and spatial resolution digital remote sensing data available at or prior to the time of the project,
and many relied to one or another degree on the use of field-based data to
calibrate models. Many of these land cover products can be leveraged to
answer a number of landscape ecological questions, which tie in closely with
planning and other decision making.

Thus, the true demand for land cover products is driven by a profound
need by society for contextual landscape information, which relates directly
to decision making. Land cover classifications and map products are customized in accordance with the decision makers' project goals or scope using a
team of image interpreters, often led by experts on the landscape elements
of interest (e.g., forestry, wetland ecology, hydrology, or geology). The team
can use the available data sources to map land cover polygons (i.e., a specific
area on the ground) according to a scope, and conduct the interpretation and
classification process to produce the desirable land cover types.

Geographic information describes the locations of landscape entities and
can be interpreted so that the spatial relationships between these entities are
understood. Most of the broad-scale geographic information produced today
resides with national and state governmental groups, but is frequently
produced at fine-to-moderate scales by local governments, individuals, cor-
porations, and other nongovernmental organizations. At one time, the need
for a computer and GIS software was a specialized occurrence; however,
with the advent of compact and powerful hardware, the processing and analyzing of these digital geographic data sets is possible for those with limited
funding (e.g., using the free and globally accessible Geographic Resources
Analysis Support System, http://grass.fbk.eu/mirrors.php, checked January
25, 2017); or using systems that are more expensive and commercially available at industry standards (e.g., Environmental Science Research Institute's
ArcGIS software, www.esri.com, checked January 25, 2017).

Categories of geographic information that cover a wide range of parameters
can then be mapped, what is referred to as a thematic map, and may include
information about topography, human population, land cover, land use,
oceans, rivers, streams, lakes, wetlands, roads, important political boundaries,
and other features of importance (e.g., national parks, monuments, and land-
marks). The level of detail described for each parameter within a thematic map
(e.g., type of vegetation: herbaceous and woody plants, or by plant species) is
dependent upon the spectral and spatial characteristics of the remote sensing
data, from which then a geospatial model is derived, as discussed above.

Although the full basis of thematic maps may not be initially understood
by all users, a typical thematic map that contains data that are digitally stored
as a series of numbers can be an excellent opportunity to explain how the
values associated with each pixel in the map can be transformed into landscape analyses. It may seem elementary to most remote sensing professionals, however, it is important to explain to the full project team (including

those less familiar with the concepts of the thematic map) that thematic maps can be thought of as checkerboards, where each grid pixel represents a data value for a particular landscape characteristic or theme (e.g., a map's topographic theme with a point elevation value and pixel value of "747," which defines that particular pixel at 747 meters above sea level). This is important to remember and articulate to the team or the decision makers in order to demystify the work and to connect with those who are familiar and interested in mathematics, so that there is a better understanding of the process behind generating thematic maps. As a GIS is used to view and measure landscape metrics or indicators, using a variety of methods, it may also be useful to explain that the overlaying of thematic maps allows several different themes to be combined in a variety of ways to extract or produce new information about combinations of initially mapped information and the spatial relationships among the themes. A good example to provide to teams or decision makers less familiar with this process is that by overlaying maps of land cover and topography, an analyst can look at the occurrence of agriculture on steep slopes, using an overlay of land cover (which includes agriculture locations) on topography (which includes elevational change, i.e., slope, across the entire landscape). These relationships can be digitally stored as a new map, which combines the information mathematically from the original set of two thematic maps. Another method to explain or demonstrate for the team or decision makers is "spatial filtering," which can be thought of as using a spatial window of a given size (e.g., a 5-pixel by 5-pixel area) to calculate values within small areas that are part of a larger map. Because spatial filtering can be used to create surface maps of metric or indicator values that help to summarize the spatial patterns of pixels' metric or indicator values in more detail than is provided, it helps to explain that this method can be applied to areas larger than the spatial window size, such as a watershed, to produce new mathematical calculations for that entire area, such as patch edge or core patch characteristics. With such examples articulated to the project team members, or decision makers, that are less familiar with spatial analyses there is a greater capability for the team to make suggestions and ask questions about the use of remote sensing to meet the landscape ecology project objectives and goals.

Because landscape ecological projects typically involve the use of several GIS data sets with multiple parameters involved, a thorough understanding of how these GIS data sets can be integrated and managed is important during the early stages of the project. With the rapid growth in GIS software and applications and the environmental scientist's capability for storing, manipulating, and visualizing data, geographic information is becoming increasingly powerful for understanding ecological data, which has shifted some of the emphasis of landscape ecology from GIS applications and manipulations (described previously) to the statistical analysis of the geospatial data. Such improved geospatial data analyses of the relationships between landscape conditions and the ecological functions of ecosystems can be enhanced by targeted (i.e., non-extensive) site-specific assessments that can then be used to generalize those fine-scale relationships across broader-scale spatial extents.

Statistical procedures can improve the understanding of the broad-scale relationships between landscape condition and the condition of ecosystems that reside therein, thereby allowing for larger data sets to be analyzed using analytical techniques, and also allowing for the inclusion of data that might otherwise cause analysis difficulties. Such geospatial statistical techniques have demonstrated initial significant relationships between ecosystem parameters (e.g., percent vegetation cover) and other mapped geographic data (e.g., road density) in the vicinity. The strengths of these relationships can be variable, and the causal relationships include a degree of uncertainty without quantification of supportive information on the ground. Geostatistical methodologies have been specifically utilized in Chapter 5 to account for environmental uncertainties and develop landscape indicators of water quality, for example. The potential limitations of using mapped geographic data to assess, for example, ephemeral ecosystems like wetlands are directly related to the capability of linking GIS-based assessments to relevant field-based assessments of wetlands (Whigham et al. 2003), an important component of determining the accuracy of landscape indicators, as a general rule.

An important development in the leveraging of remote sensing data are *vegetation indices,* which are derived from spectral reflectance values and based on the differential absorption and reflectance of energy by vegetation in the red and near-infrared portion of the electromagnetic spectrum (Thenkabail et al. 2009, 2012). In general, green vegetation absorbs energy in the red region and is highly reflective in the near-infrared region (Lillesand et al. 2014). A number of vegetation indices have been formulated and utilized for monitoring vegetation change. Of these vegetation indices, the Normalized Differential Vegetation Index (NDVI) has been used most widely for monitoring terrestrial vegetation dynamics (Luo et al. 2013; Balzarolo et al. 2016). The NDVI compensates for some radiometric differences between or among images. The differences in NDVI values of two images in certain cases responds to changes in land cover. Singh (1989) concluded that NDVI differencing was among the most accurate of change detection techniques, and many indices have been developed and tested that demonstrate similar utility for the production of vegetational land cover products (Thenkabail et al. 2009, 2012).

2.7 Leveraging Digital Data for Image Processing

Because most of the remote sensing data that is used on a typical landscape ecology project is digital, image processing and modeling of those collected data streams using a GIS is a distinct time saving and analytical advantage (e.g., Maidment and Djokic 2000; Maidment 2002; Lyon 2003; Madden 2015), which we make use of throughout the projects described in this book. Imagine the costly and time-consuming alternative for the vast area of the Great Lakes Basin, using manual photo interpretation methodologies! The advent of digital image processing has added great analytical power to evaluations, including the detection of change, and specialized ecosystem features such as plant taxa, ephemeral wetlands, and plant water stress. Further, the availability of fine resolution satellite data with very good repeat coverage has brought a whole new dynamic to the effort. Consequently, remote assessments are becoming more and more sophisticated and detailed in their outputs. We present some of these approaches in the remainder of this book, which supply a great deal of unique and powerful capability, and ideas of the options you may need, for implementation in landscape ecology projects that cover large-to-vast areas of the landscape.

Because wetlands are transitional ecosystems, containing both upland and aquatic elements at different times and sometimes a mixture of both at a single instance, they are extremely difficult to assess remotely, and consequently many of the advances in remote sensing for landscape ecology has occurred in these challenging ecosystem environments, both for field analysts and remote sensing analysts. Accordingly, much of the groundbreaking work in these challenging environments has been developed for the efficient and accurate evaluation of land cover change where wetlands are either embedded within other land cover types, or directly adjacent to other land cover types, from both airborne and satellite remote sensing data (e.g., Buzzelli 2009; Adam et al. 2010; Klemas 2011; Homer et al. 2015). Accordingly, wetland ecologists, landscape ecologists, and remotes sensing scientists/practitioners have worked hand-in-hand on innovative procedures that were entirely developed on wetland ecosystems in North America, South America (including the trans-Amazon), and Africa for purposes of evaluating condition, often including the potential influences of climate change (Thenkabail et al. 2005; Velpuri et al. 2009).

Due to the importance of change detection in research and management missions (e.g., Johansen et al. 2010; Pollard et al. 2010), large efforts have also been devoted to the development of these innovative remote sensing and landscape ecology procedures for operational evaluations of landscape and ecosystem change, using satellite data (Ramsey and Jensen 1995; Ramsey et al. 1998; Thenkabail et al. 2012). Early development of broad-scale assessment of land cover led to more regional scale studies, such as change detection of land cover in Mexico across three decades, as part of the North American Landscape Characterization (NALC) project (Lunetta et al. 1993, 1998; Lyon et al. 1998; Vicente-Serrano et al. 2008), which is a seminal piece of work that provided early insight into challenges of utilizing multiple satellite images to produce a national coverage of land cover.

The National Land Cover Database (NLCD) carried forward the ground-breaking work of NALC, and provided similar Landsat-based, 30-meter resolution, land cover information for the United States. The NLCD continues to provide spatial reference and other important descriptive data for the characteristics of land surfaces, which includes urban, agriculture, forest, percent impervious surface, and percent tree canopy cover. It has supported numerous federal, state, local, and nongovernmental programs and specific applications and provided the public with the opportunity to evaluate ecosystem status and condition, improving users' capacity to:

- Understand spatial patterns of biological diversity
- Address climate change and associated outcomes
- Analyze ecosystem/ecological services and land health
- Develop land management policies and procedures in the context of land cover type and configurations (Vogelmann et al. 2001)

NLCD products are developed and produced by the Multi-Resolution Land Characteristics (MRLC) Consortium, which has continued as a multidisciplinary partnership of federal agencies led by the USGS. All NLCD data products are now freely available and downloadable from the MRLC Internet website: http://www.mrlc.gov (checked January 25, 2017), and products can be viewed on the MRLC Consortium viewer at the same site.

A measure of the value of MRLC products was the second set of content developed in the last decade, referred to as MRLC 2006 (Foody 2002; Vogelmann et al. 2010; Homer et al. 2015). The net result of this leadership for more than four decades is the availability, at no cost, of land cover themes over several decades and at several spatial resolutions. The user can utilize these data sets to develop regional or local analyses of ecosystems. The availability of the Landsat data archive, at no cost from the USGS, allows for deeper analyses or temporal studies (Frohn et al. 2009). These resources are a great advantage to the new users of land cover data as well as the more seasoned users of remote sensing data (Na et al. 2010).

There are a number of approaches and methods for developing remote sensing data for the detection of landscape change over time. Often users wish to develop their own data sets from raw remote sensing data as opposed to using available land cover data, such as MRLC. The process begins with selecting multiple date data sets to identify the desired features, ecosystem, or landscape of interest, and assembling the chronology of data, and then analyzing for time steps that may reveal a change in the desired features, ecosystem, or landscape, using image processing techniques. This is the process of remote sensing change analysis, or change detection. There are a number of complicating factors that may make change detection more challenging, including lack of sufficient data coverage to resolve the phenomenon of interest (either spatially or temporally), or cloud cover or other obscuring factors. Currently, the use of twice-daily MODIS coverages advances this type of global endeavor and adds greater potential for monitoring plant phenology, vegetation indexes, and a host of other outputs for the use in landscape ecology, which can successfully support extremely broad-scale analyses of landscape change (https://lpdaac.usgs.gov/dataset_discovery/modis/modis _products_table, checked January 25, 2017).

2.8 The Importance of Preprocessing

By the time one can use a set of data for the leveraged uses described above, a number of processing procedures have already likely occurred, which add value and depth to the raw data collected at the sensor. These data processing steps that result in high-quality imagery for the further derivation of classified outputs, indexed images, and categorization of land cover types are important and are briefly considered here. A large goal of applying these preprocessing algorithms to the data is to produce images with less noise and more signal from the original sensor collection. All or some of these preprocessing procedures may apply to a given image and the selective use of each is dictated by the type of data and by the objectives of the project. The procedures include smoothing of spectral variability, image compositing to remove clouds, image enhancements such as ratioing and indices, and more.

Processed images will often be smoothed or deconvoluted by de-striping of the images. This procedure removes the variability from rows of data in a digital image, which may be a result of differential input responses at the sensor or some other radiometric problems. Often there will be a general deconvolution of processed images to correct these radiometric problems, resulting in resampling pixel size, perhaps from a finer resolution to a coarser resolution pixel size, to maintain geometric fidelity between or among images that are being combined or compared.

Special efforts have been made over the years to develop data processing methods to create cloud-free composite images, using multiple dates of images of the same area (Loveland and Ohlen 1993; Chander et al. 2009). This processing is necessary when a full ground coverage is needed in areas with a lot of cloud cover, such as high altitude mountain or tropical regions. This is a very typical need when utilizing large satellite-based data collection efforts, such as projects and programs at major data collection and compilation centers, such as the USGS EROS Data Center. An early example of this work was that applied in support of the Advanced Very High Resolution Radiometer (AVHRR) biweekly vegetation index composites program over the years (Young and Wang 2001; Vicente-Serrano et al. 2008; Vogelmann et al. 2010), and later with similar programs that utilize MODIS data (Linderman et al. 2010; Lunetta et al. 2010). Such composite images are made by spectrally filtering clouds from multiple overlapping images and then combining cloud-free portions among multiple images, usually from different dates, thus "removing" those cloudy or cloud-shadowed areas. Composite images often then undergo additional processing to adjust the multiple images spectrally. Corrections of the scenes to a set solar zenith angle is one approach used to establish similar solar illumination conditions among merged images (Schott 2007; Chander et al. 2009) or other image equalization procedures (Lunetta and Elvidge 1998; Thenkabail et al. 2009; Vogelmann et al. 2010).

Because spatial misregistration of images can tend to reduce the overall accuracy of any digital change detection effort, single of mosaiced imagery must be registered to a valid reference. The effects of misregistration are most severely observable and measurable for projects that rely upon change detection techniques. For change detection project data, digital coregistration among images, accuracies on the order of half a pixel or less are necessary to create data sets that can be analyzed for change with minimal errors (Lunetta et al. 1993).

2.9 The Role of Classification Systems

The classes in a preliminary land cover image can be labeled using standardized land cover classification systems (Jensen 2004). Classification systems were developed to specifically support project objectives and to be compatible among different maps. Therefore, most classifications can be "cross-walked" to one or another extent with the other major land cover classification systems, by either finding analogue land cover type classes or by lumping and splitting between two different classification systems so that they become the same or similar.

Any customized land cover classification system should always be designed to be compatible with those that are widely accepted and utilized (i.e., Anderson et al. 1976; Brown et al. 1979; Cowardin et al. 1979) or are compliant with another accepted system that is relevant to project goals and the user's needs. Most land cover classification systems have been optimized for inventory types of applications, and include most of the fundamental land cover types that one would utilize in a project. It is recommended that any project that develops a unique classification system do so with the understanding that comparisons with other landscape data sets may be difficult and make comparisons, modeling, or other analyses with those areas challenging.

The assignment of land cover class types to a preliminary land cover product should be performed by personnel familiar with the traditionally developed systems and standard techniques, along with the possible assistance of local partners, especially those individuals familiar with the land cover types, features, and other information and activities in the project area. Naturally, this sort of work is also greatly facilitated by field work and supporting information such as other project results, research studies, soil surveys, crop information, topographic information, digital elevation models, or other land cover information or imagery.

The level of detail of the classification system chosen should be selected by the project's working groups, and is a decision dependent upon the needed specificity of the classes used. Classification Level 2 or Level 3 is often used for typical land cover type detail (Anderson et al. 1976; http://landcover.usgs.gov/pdf/anderson.pdf, checked January 25, 2017) because this approach strikes a good balance between (1) provision of sufficient detail and (2) accounting for the technical capabilities of the specific remote sensing data used. It is recognized that differentiation between classes at Level 2 will be difficult and in some cases not possible without the aid of high resolution and specialized remote sensing data, such as RADAR, LiDAR, or Shuttle topography data (Henderson and Lewis 2008; Lunetta et al. 2010; Abood et al. 2012).

2.10 Change Analysis

Much of the historical work on landscape conditions has been focused on combining statistics from agricultural, natural resource, and conservation inventories that have already occurred. In the United States, reports often describe historical numbers at the time of the U.S. Public Lands Survey, or as homesteaders moved lands into agricultural or range production. Such inventory statistics had been retained by those agencies in support of agriculture programs and, for example, include acreage of forested areas harvested or acreage of wetlands drained and placed into agricultural production (Figure 2.5). These assessments are useful for an initial understanding of the loss of previously intact ecosystems, such as forest or wetland ecosystems. In some cases, for trends in amounts of total forested areas and wetlands the values tend to be broad in definition and in areal coverage (Dahl and Johnson 1991; Dahl 2006; Dahl and Watmough 2007).

Evaluations of ecosystems and their change over time have also been completed using aerial photographic or image products and interpretations (Lyon 1993, 2003; Adam et al. 2010; Lyon and Lyon 2011; Schlesinger et al. 2015). Evaluations of multiple dates of historical images allow for interpretation of conditions and changes over time (Garofalo 2003; Schmidt et al. 2015). It is recommended that the use of four or more dates of images leads to best results in change analyses and, if possible, particularly when landscape change over time is rapid or stochastic among elements in images (Schott 2007; Lyon and Lyon 2011). The abundance of imagery in archives, currently acquired imagery, and methods to integrate these data for landscape change analyses adds depth to historical or inventory projects and, depending upon the number of dates available, can provide a very illustrative example of the timing of landscape changes, which are quite demonstrative of the history of a landscape and often resonate well with a multitude of audiences in a specific locality or region. From a number of dates of historic photographs or

FIGURE 2.5
Conversion of land from natural land cover types to human-engineered landscapes is common throughout the world, and remote sensing data and analyses of these patterns is important to better understand how landscape processes may be changed from these landscape configuration changes.

images, hydrological, vegetational, soil condition, or human activity data can be interpreted effectively. Evaluation of the presence or absence of hydrological conditions or stressors over time are also very demonstrable by these techniques (Federal Geographic Data Committee 1992, 2008, 2010; Lyon 1993; Lyon and Lyon 2012). Other efforts where remote sensing products have been used for the analyses of landscape change are numerous in the literature (e.g., Lee and Marsh 1995; Schaal 1995; Lunetta and Elvidge 1998; Garofalo 2003; Vicente-Serrano et al. 2008; Lunetta et al. 2010; Meyer et al. 2015), and all demonstrate one of the true utilities and needs for repeated collection of remote sensing in natural areas. Post-categorization change detection involves the categorization or computer classification of composited images, as described earlier in this chapter, and labeling of land cover thematic classes from each year using the same class types from the same classification systems (Bolstad and Lillesand 1992; Villeneuve 2005; Baker et al. 2006). The locations of change can then be identified by the areas of change in land cover from the earlier date image, and compared to the latter date image.

2.11 Assessment of Accuracy

Assessment of the accuracy of categorized or classified remote sensing products is a final and necessary step to ensuring that a quality product for subsequent analysis is the output of a project and that the products meet quality assurance and quality control (QA/QC) objectives. This can be accomplished by field work or image interpretation of the land cover classifications, and conducted according to the scope or guidelines of a team of image interpreters or field personnel (e.g., Sinha et al. 2012). The image interpretation/field team should utilize an accepted land cover classification system as specified in the original scope of the project to verify the results of the product (e.g., a categorization of land cover).

Image interpreters should utilize a number of other sources of data, which many studies utilize. The data sources can include color infrared high altitude photographs, Landsat, SPOT, fine resolution satellite data such as GeoEye Image data, aerial image and map atlases, or USGS quadrangle maps and digital files. In addition to these sources, the team should utilize any other sources available in the locality or region that may be unique or specialized, which may often rely on nontraditional information, such as inventories and sociological data, to name a few. These unique or specialized data are often in formats that may need to be transferred to digital formats to allow for compiling and manipulation along with the other more traditional data sets involved in the project, and often ensure quality of the project outputs by truthing some of the assumptions with other independent sources of information. These additional data include USDA Natural Resource Conservation Service soil surveys, U.S. Fish and Wildlife Service National Wetlands Inventory (NWI 2012) maps/data, available USGS orthophoto quadrangle products, and maps of federal property boundaries.

QA/QC work can be effectively accomplished by utilizing complete or partial county coverages for the area assessed, where there can be archival photographs or other fine-scale imagery that can be used along with other data sources to map and check land cover accordingly. These coverages are often at medium and low altitude, and they supply sufficient detail that can support interpretations from higher altitude photographs or satellite products. These photographs or images represent a very valuable resource that local groups bring to the project, and the use of these detailed and contemporary photos will both speed the classification work and enhance the accuracy of thematic mapping. Computer categorizations can also be conducted in a few local areas, using an independent data source from those used for the original thematic map. These quality determinations should also be made after the final digital databases or land cover products are produced. In this manner, the final polygons and classifications in any project can be independently checked against original photographs, supporting data sources, or the original interpreted data.

Field-based validation of the landscape metrics/indicators is also very much encouraged as a QA/QC step in a project's implementation, and a vital (often neglected) step that requires a statistically sound methodology prior to entering the field to conduct sampling. A typical way of determining if a landscape metric/indicator is sensitive is to hold it to a predetermined standard of acceptability (e.g., for a linear regression or ANOVA, a significance level of α = 0.05), but this standard is not always the same for every metric, geography, ecosystem, and circumstance or audience. This step relies on the a priori knowledge and expertise of the research team and is a standard method for designing a hypothesis test. Thus, a critical analysis step is for the project team to agree on the level of sensitivity that is required to satisfy the various project or policy goals, prior to the commencement of field validation.

2.12 Brief Evaluation of Landscape Metrics

There is an amazingly wide variety of landscape metrics available to a team to utilize in the course of project development and implementation, and many software programs are available to calculate them (O'Neill et al. 1988; Turner and Gardner 1991; Baker and Cai 1992; McGarigal and Marks et al. 1994; Riitters et al. 1995). Many of these landscape metrics have been shown to be highly correlated with one another as well (Riitters et al. 1995). One of the lessons of the work outlined in subsequent chapters is that the type of metric chosen is dependent upon the needs of the user, and an understanding of the ecological significance of trends of the metrics is important to master, rather than selecting a metric that is "popular" or choosing one that worked well in some other geography or circumstance. In a factor analysis study, Riitters et al. (1995) analyzed 55 typically utilized landscape metrics for their statistical independence. Their conclusions provide us with a valuable categorization and perspectives on the six fundamental landscape metric types that we will encounter in Chapters 4 and 5, and in all of the landscape ecological work that you will encounter in your circumstances, regardless of their origin or specific calculation (listed below in alphabetical order):

- Average patch perimeter/area ratio
- Average patch perimeter/area ratio orthogonally adjusted
- Contagion
- Dominance
- Fractal Dimension from perimeter/area
- Number of classes

The U.S. Environmental Protection Agency (USEPA) also ranks the status of a number of landscape metrics into one of three categories based upon their predominant uses:

1. For use as watershed integrity indicators (Table 2.1)
2. Landscape stability and resilience indicators (Table 2.2)
3. Biotic integrity and diversity indicators (Table 2.3)

TABLE 2.1

Ranking of Watershed Integrity Indicators

Indicator Name	Rank
Contagion	C
Fractal Dimension	C
Dominance	C
Lacunarity	A
Erosion risk	A
Flood indicator	A
Riparian zones	C
Loss of wetlands	C
Agriculture near water	B
Kilometers of new roads	B
Amounts of agriculture and urban	C
Watershed/water quality indicator	A

Source: USEPA, Landscape Monitoring and Assessment Research Plan, EPA 620/R–94/009. Office of Research and Development, Washington, DC, 1994.

Note: A = requiring further conceptual development; B = requiring testing for feasibility/sensitivity; C = ready for field tests and implementation.

TABLE 2.2

Ranking of Landscape Stability and Resilience Indicators

Indicator	Rank
Contagion	C
Fractal Dimension	C
Dominance	C
Lacunarity	A
Diffusion rates	A
Percolation backbone	B
Percolation thresholds	B
Kilometers of roads	B
Recovery time	A
Land cover transition matrix	A

Source: USEPA, Landscape Monitoring and Assessment Research Plan, EPA 620/R–94/009. Office of Research and Development, Washington, DC, 1994.

Note: A = requiring further conceptual development; B = requiring testing for feasibility/sensitivity; C = ready for field tests and implementation.

TABLE 2.3

Ranking of Biotic Integrity and Diversity Indicators

Indicator	Rank
Contagion	C
Fractal Dimension	C
Dominance	C
Lacunarity	A
Change of habitat	C
Habitat for endangered species	C
Loss of rare land cover	C
Corridors between patches	B
Amount of edges	C
Edge amount per patch size	B
Patch size distribution	C
Largest patch	B
Interpatch distances	B
Linear configurations	A
Actual vs. potential vegetation	B
Wildlife potential	A
Kilometers of new roads	B
Diffusion rates	A
Percolation backbone	B
Percolation thresholds	B
Resource utilization scale	B
Scales of pattern	A
Cellular automata	A
Pixel transitions	A

Source: USEPA, Landscape Monitoring and Assessment Research Plan, EPA 620/R–94/009. Office of Research and Development, Washington, DC, 1994.

Note: A = requiring further conceptual development; B = requiring testing for feasibility/sensitivity; C = ready for field tests and implementation.

The three rankings employed by the USEPA relate to the metric's relative readiness for use by personnel in their work in 1994 (USEPA 1994). Those rankings are relevant today, and are as follows:

A: Requiring further conceptual development

B: Requiring further testing for feasibility/sensitivity

C: Ready for field tests and implementation

Although some time has passed since the initial evaluation, above, of the Contagion and Fractal Dimension metric types, they both continue to be the type of metrics that are well developed and understood from a theoretical ecological perspective, and are certainly among those metrics that are ready for field tests and implementation. The USEPA (1994) also concluded that change in landscape patterns could be well characterized by the three metric types/categories of Contagion, Fractal Dimension, and Dominance. Accordingly, landscape change can be analyzed by calculation of a three-dimensional Euclidean distance utilizing all three metric categories, as follows:

$$\text{Change} = ((X1 - X2)^2 + (Y1 - Y2)^2 + (Z1 - Z2)^2)^{1/2} \tag{2.1}$$

where, in this example, X is Dominance, Y is Contagion, and Z is Fractal Dimension; and at some magnitude (as yet unknown), this shift would represent a phase change in the landscape (USEPA 1994).

The Contagion metric was first proposed by O'Neill et al. (1988) and later by several others (Turner and Ruscher 1988; Turner 1989, 1990a, 1990b; Graham et al. 1991; Gustafson and Parker 1992; Li and Reynolds 1993; USEPA 1994, 1996) as a measure of clumping or aggregation of patches. It is also used as an indication of the degree of fragmentation of a landscape. Contagion is described as:

$$\text{Contagion} = \frac{2\ln(t)\sum_{i=1}^{t}\sum_{j=1}^{t}((n_{ij}/N))\ln((n_{ij}/N))}{2\ln(t)} \tag{2.2}$$

where n_{ij} is the number of shared pixel edges between classes i and j and N is twice the number of total pixel edges since there is double counting of edges (e.g., AB and BA edge are counted twice) and t is the total number of classes. Twice the natural log of t reaches its maximum when all pixel edges of classes i and j have the same proportion. In such a case, a given pixel of one land cover type would have an equal chance of being adjacent to another pixel of any land cover type. Division by twice the natural log of t normalizes the value of C between 0.0 and 1.0. Theoretically, at lower values of C, there are many small patches, and thus the proportion of pixels being adjacent to a given land cover type are nearly equal. Accordingly, as Contagion approaches 1, there are large contiguous patches on the landscape (O'Neill et al. 1988; Li and Reynolds 1993).

The Contagion metric has been used extensively in the analysis of landscapes and ecosystems. Contagion has been used to detect changes in spatial patterns of clumping across a variety of landscapes in the United States (O'Neill et al. 1988; Turner and Ruscher 1988; Turner 1990a, 1990b; USEPA 1994, 1996; Riitters et al. 1995) and the Brazilian Amazon (Dale et al. 1993, 1994). Contagion has been used to detect the effects of spatial scale on landscape patterns (Turner 1989). Some examples of the use of Contagion include the analysis of the relationship between land cover proportions and spatial pattern (Gustafson and Parker 1992), the use of Contagion in ecological risk assessment (Graham et al. 1991), and the use of Contagion as a measure of image texture (Musick and Grover 1991).

Despite its widespread use, Contagion may be adversely influenced by a number of problems that have not been thoroughly addressed with respect to remote sensing data. The following is a list of potential problems with the Contagion metric:

1. Effects of measurement resolution on Contagion
2. Effects of raster orientation on Contagion
3. Effects of varying the number of land cover classes on Contagion

Fractal Dimension has been used for measurement, simulation, and as a spatial analytic tool in the mapping sciences. Fractal Dimension has also been used to characterize landscape complexity (Krummel et al. 1987; O'Neill et al. 1988; De Cola 1989; Lam 1990). The term *fractal* was introduced by Mandelbrot (1977) and its range of applications later developed and expanded to other disciplines (Lam and Quattrochi 1992). An overview of fractals can be found in Goodchild and Mark (1987).

The use of fractals in remote sensing is relatively new (Lam and Quattrochi 1992). Several studies have used fractal analysis with remote sensing images (De Cola 1989; Lam 1990; Agnon and Stiassnie 1991; Linnet et al. 1991; Meltzer and Hastings 1992). Lam (1990) found that different land cover types had different Fractal Dimensions for each band of a Landsat TM scene. Lam (1990) also maintained that the use of fractals on individual bands may serve as guidelines in the future for selecting bands for display and analysis.

Changes in Fractal Dimensions in remote sensing images have implications for changes in environmental conditions (Lam and Quattrochi 1992). A number of studies have found that the Fractal Dimension of the landscape varies according to the type of land use (O'Neill et al. 1988; De Cola 1989). For example, forest areas tend to have more complex shapes, and thus manifest high Fractal Dimensions, while agricultural areas tend to have simple shapes, and thus have low Fractal Dimensions. There also appears to be a correlation between Fractal Dimension and the degree of human disturbance of the landscape (Krummel et al. 1987; O'Neill et al. 1988; Turner and Ruscher 1988; De Cola 1989). As human disturbance increases (e.g., increased intensity of cultivation), the Fractal Dimension of the landscape decreases (Krummel et al. 1987; O'Neill et al. 1988).

There are a number of methods that can be used to determine Fractal Dimension of a landscape (Lam and Quattrochi 1992). A common method of determining Fractal Dimension on images in landscape ecology is based on perimeter-to-area relations, with several examples of this work in subsequent chapters. Fractal Dimension as originally derived by Mandelbrot (1977) describes a scale invariant power relationship between perimeter and area in the form:

$$P = KA^{D/2} \tag{2.3}$$

where P = perimeter, A = area, D = Fractal Dimension, and K = constant of proportionality. Fractal Dimension, D, may then be determined from Equation 2.3 as:

$$\ln P = \ln(kA^{D/2}) \tag{2.4}$$

$$\ln P = \ln k + \ln(A^{D/2}) \tag{2.5}$$

$$\ln P = \ln k + (D/2)*(\ln A) \tag{2.6}$$

$$D = 2*(\ln(P) - \ln(k))/\ln(A) \tag{2.7}$$

The value of D is between 1 (with simplest shapes) and 2 (most complex shapes) (O'Neill et al. 1988). Landscape ecologists have used the perimeter-to-area relationship with map and image data to characterize the complexity of landscape patterns (Krummel et al. 1987; O'Neill et al. 1988). The Fractal Dimension of landscapes has been most commonly estimated by regression methods (Krummel et al. 1987; O'Neill et al. 1988; Milne 1991; Baker and Cai 1992; McGarigal and Marks 1994). Referring to Equation 2.6, the natural logarithm of perimeter is regressed against the natural logarithm of area for all patches on the landscape. The $\ln(k)$ is the y-intercept and for squares k = 4. Fractal Dimension is estimated as two times the slope of the regression since the slope is $D/2$.

Although the result has been referred to as Fractal Dimension, this is an inappropriate use of terminology both in conceptual foundation and technical implementation. The presupposition in this use of Fractal Dimension is that self-similar patterns exist across various sizes of landscape patches, which are in turn much larger than the measurement resolution. Fractal Dimension is a measurement across scales which is predicated on the concept of actual or statistical self-similarity across measurement scales. In other words, patches of different sizes at one scale are used as a surrogate for a change in scale. Furthermore, unless the constant of proportionality (k) is known, there are two unknowns in Equation 2.7. Regression takes care of the unknown $\ln(k)$ by determining the y-intercept. The constant of proportionality (k) is only known for geometric equilateral shapes (e.g., a square) and true fractals (e.g., a Koch curve). The value of k is the constant that relates the perimeter of a shape to the square root of the area of a shape. For instance, for a square $P = 4 * A^{1/2}$; thus, $k = 4$. For other geometric equilateral shapes $P = k * A^{1/2}$ and for fractal shapes $P = kA^{D/2}$ (Equation 2.3). The question remains whether the $\ln(k)$ (i.e., y-intercept) is estimated correctly for other shapes during regression in estimating Fractal Dimension.

The trend toward the use of Fractal Dimension as a general index of landscape pattern or complexity is evidenced by the number of studies that have utilized it as a metric. The majority of studies for the use of Fractal Dimension using linear regression in landscape ecology concerns changes in land cover patterns from broad scales (Gardner et al. 1987; Lathrop and Peterson 1992; Riitters et al. 1995) to regional scales (Iverson 1988; Turner and Ruscher 1988) to local scales (Pastor and Broschart 1990; Rex and Malanson 1990; Mladenoff et al. 1993). Sugihara and May (1990) showed a number of applications of Fractal Dimension in ecology. Gustafson and Parker (1992) used Fractal Dimension to examine the relationship of land cover proportion to landscape pattern. Dale et al. (1993, 1994) used Fractal Dimension to characterize simulated spatial patterns in the Brazilian Amazon. The importance of Fractal Dimension is further shown by the EPA recommendations that it is a technique that is ready for field tests and implementation in its Environmental Monitoring and Assessment Program (EMAP; USEPA 1994, 1996).

The Fractal Dimension metric has embedded in it a fundamental assumption that there is a power law relationship between perimeter and area. The following is a list of potential problems with Fractal Dimension as a landscape metric of shape complexity:

1. Problems with perimeter/area regression
2. Problems of Fractal Dimension with the raster data structure
3. Effects of measurement resolution on Fractal Dimension

Landscape ecologists have long used perimeter-to-area relationships with map and image data to characterize the complexity of landscape patterns (Krummel et al. 1987; O'Neill et al. 1988). The Fractal Dimension of landscapes has been most commonly estimated by regression methods (Krummel et al. 1987; O'Neill et al. 1988; Milne 1991; Baker and Cai 1992; McGarigal and Marks et al. 1994). Referring to Equation 2.6, the natural logarithm of perimeter is regressed against the natural logarithm of area for all patches on the landscape. The $\ln(k)$ is the y-intercept and for squares $k = 4$. Fractal Dimension is estimated as two times the slope of the regression since the slope is $D/2$.

It is important to recognize that the conclusions of the work of USEPA, as summarized by Frohn (1997), has been used as the basis for the past 20 years of landscape ecological analyses, including the work outlined in the subsequent chapters of this book, utilizing the theory and practice of many professionals across thousands of landscapes and in tens of thousands of eco-systems, and under a variety of circumstances derived from societal needs. The several case studies provided in this book outline that same diversity of approach and need, and follow the general guidelines of the above cat-egorizations and priorities, with modifications as it is appropriate, given data availability and needs of the project. It is critical to understand that the summary information in this section is intended to serve as a reference for understanding the "full universe" of possibilities for the use of landscape metric types; however, the selection and development of metrics for specific project needs may vary, as in the examples provided in this book.

2.13 Information Management Factors

There are a variety of information needs and priorities in landscape charac-terization and ecosystem management, dominantly driven by societal, pro-fessional, and scientific needs and processes. Increasingly, there is a need to develop landscape ecological projects along with communities and other stakeholders, and this need has resulted in numerous examples of excellent collaborative initiatives at a variety of scales, such as the globally focused International Union for the Conservation of Nature (https://www.iucn.org, checked January 25, 2017); a nationally focused Collaborative Forest Landscape Restoration Program, developed by the U.S. Forest Service (https://www .fs.fed.us/restoration/CFLRP/overview.shtml, checked January 25, 2017); and a locally focused Resilient Landscapes Program, developed by the San Francisco Estuary Institute and the Aquatic Science Center (http://www.sfei .org/he, checked January 25, 2017). These highly focused and purposefully collaborative examples are in part a result of a variety of issues they are designed to address, which require the melding of science, culture, and the basic knowledge of the landscape needed to accomplish the goals within the study areas. Traditional issues and questions encountered in any landscape understandably come from the operational information needs of community members, conservation groups of a region, and governmental organizations in that region. The specific data and information needs are also inclusive of the characteristics of the landscape, such as biophysical feature types and locations, ecosystem exposure and risk characteristics, condition of a site and facilities or resources, management activities, and the cultural resources of the area. The domain of all of these issues often stems from both regulatory requirements and community engagement requirements or expectations, particularly when federal, state, and local agencies are involved. It is more and more typical that these requirements emanate from different questions and needs from the public as a project develops.

Through increasingly more involvement and comanagement of landscape ecology projects with community members and decision making groups, early on, in the development of a project, new types of information and ideas about how project information can be reported and interpreted are necessary to highlight. This type of inclusion of diverse project team members, with community or decision making leaders, allows for better access to new information from local perspectives and sources as well, which benefits the project.

Some of the data needs of a diverse project team include data on human use, data for identification and management of a variety of ecosystems, cultural features, databases to support management operations and maintenance, and data to support public inquiry and response systems. Although some of these data may be in nontraditional formats (e.g., oral or in nonstandard written records), or qualitative in nature, such information can help to guide the landscape ecology project team toward other more formal or broad-scale databases, or supplement them, potentially leading to a melding of both quantitative and qualitative approaches that satisfy all of the team members' needs.

2.14 Management Uses of Landscape Metrics and Indicators

Many of the questions related to assessing landscape condition and measuring ecosystem functions may involve the encroachment in and around ecosystems by human land uses, such as urbanization or agriculture. There are a number of imaging and GIS methods to identify the location and the probability that a disturbance or exposure has occurred, or is destined to impinge upon the ecosystem, directly and physically, or indirectly through secondary factors like hydrologic alteration or a loss of habitat connectivity. Typically, a disturbance or exposure can be identified by the change in digital or analogue remote sensing imagery and a comparison with multiple images allows a thorough assessment of condition changes over time. Once a disturbance or exposure has been identified it can be verified with field work, typically utilizing relatively limited sample points across the landscape. This field sampling approach allows for the efficient use of personnel by guiding them to a probable activity, and verifying that activity, and thereby optimizing for field team safety and gathering of important and detailed field-based information. This combined remote sensing and field-based approach represents a good application of the capabilities of both ecology and technology, which actively puts the remote sensing and data processing technology in the role of supplying likely locations, and identification, of probable disturbances or exposures, and allowing human-directed field checks to verify disturbances or exposures, with certainty.

Utilizing field-based approaches, remote sensing, and GIS technologies together can provide a valuable method for assisting in the prioritization of field crew work, leading them to hypothesized field-based elements, which can be verified physically. Such information can also, importantly, aid in the dispatch of field crews and in the facilitation of natural resource managers' evaluations of a site. It should be recognized that these combined field-based and remote sensing detection methods are particularly cost effective when compared with the relative costs of traditional methods of field work, and associated procurement (Falkner and Morgan 2001). It is these and other technical efficiencies that make remote sensing–based landscape ecological projects so important to the needs of real people, those who live in communities across landscapes that are to be assessed. These societally important needs and issues are explored in additional detail in the following chapter, as a specific example of the interdependence of the landscape and the people that reside there (i.e., the U.S. and Canada Great Lakes Region). As we transition to a thorough discussion of practical applications of using remote sensing for landscape in ecology in Chapters 4 and 5, the next chapter outlines the importance of context, depth, and powerful outcomes of this work, made possible by the theoretical and technical aspects outlined in this chapter.

3

Utilizing Remote Sensing to Solve Landscape Ecology Challenges in Diverse Landscapes

This chapter focuses in depth on the details of how, within a biophysically and societally complex landscape, the utilization of remote sensing for landscape ecology is possible, and indeed provides an excellent set of tools for:

1. Improving the understanding and application of data within a variety of societal contexts
2. Successfully addressing specific regional needs of decision makers
3. Successfully applying the fundamentals of remote sensing and landscape ecology

The Laurentian Great Lakes (United States and Canada) is the example used in this chapter to demonstrate a successful resolution to the above three needs, which has resulted in a major success for the region on many levels of knowledge, science, and society; special discussion of these aspects has been reserved for the Great Lakes project, in particular. The Laurentian Great Lakes reside within a large and complex landscape that has long been a focus of necessary ecological restoration. The landscape ecology metrics designed for the Great Lakes in Chapter 4 provide the specific technical aspects of the project, and owing to the relative importance of the Great Lakes region and the number of societal pressures involved, it also provides an excellent example of foundational landscape scale issues, as well as a strong public interest in restoring the ecology of the region. The focused discussion in this chapter is meant to additionally supplement the application outputs in Chapter 4 for the Great Lakes, and outline specific approaches recommended in Chapter 2, so that you can see how to step forward from the foundational principles of remote sensing for landscape ecology to the practical "real world" set of problems that are typical of many landscape scale projects.

The Great Lakes work described in this book as a whole immediately preceded the expansive and innovative Great Lakes Restoration Initiative (GLRI), initiated by the U.S. Congress in 2010 with an emphasis on ensuring the restoration of the Great Lakes, the largest freshwater ecosystem on the planet. The GLRI was designed to provide additional assets to gain ground on years of work toward the long-term objectives of restoring the highly urbanized, industrialized, and agricultural international watershed, and enhance ecological research for the purpose of improving the condition of ecosystems, as well as the well-being of those living in the Great Lakes Basin. The GLRI has been an impetus for exceptional government and science coordination, through the Interagency Task Force and the Regional Working Group, which are coordinated by the U.S. Environmental Protection Agency. The integration of the GLRI work and mission has led to an unprecedentedly successful list of accomplishments, which are substantially served by remote sensing data, field data, and other information on the impressive list of projects and results (https://www.glri.us, checked January 25, 2017).

The GLRI work has subsidized the cleanup activities required to delist five Great Lakes Areas of Concern and to formally delist the Presque Isle Bay Area of Concern, a very noteworthy change from the 25-year period before the initiative, during which only a single Area of Concern was delisted. The GLRI work has been utilized to support more than 2,000 activities to enhance water quality, to ensure and reestablish local natural surroundings and species, to counteract and control invasive species, and to address other Great Lakes landscape ecological issues.

Fiscal Year 2015–2019 plans for the GLRI include an emphasis on the greatest dangers to the Great Lakes and builds upon earlier work with non-federal collaborators to execute additional restoration ventures. To guide this work, government organizations have drafted the GLRI Action Plan II, which lays out a specified period of work on Great Lakes ecological issues and related human well-being issues, large portions of which will take decades to determine, in order to meet the ultimate goal of meeting U.S. responsibilities under the United States–Canada Great Lakes Water Quality Agreement. The GLRI Action Plan II utilizes an approach that subsidizes research and other project work, which ensures a reestablishment of the largest surface water framework on the planet that focuses on cleaning up Great Lakes Areas of Concern; preventing and controlling invasive species; reducing sedimentation, erosion, and sediment loading to water bodies, which contributes to harmful algal blooms; and restoring ecosystem conditions.

3.1 Overview of Project Approach

A landscape approach was used in the Great Lakes Basin project because the interdisciplinary science of landscape ecology is ideal for examining the distribution of ecological communities or ecosystems, the ecological processes that affect those patterns, and changes in both the patterns and processes over space and time (USEPA 2001a), particularly for such a vast area. Because the broader context of land and conditions surrounding the ecological communities or ecosystems of interest is "the landscape" in this example, this larger area serves as both a conceptual unit for the study of spatial patterns in the physical environment and the influence of these patterns on important environmental resources across the vast area around the Great Lakes, and also a practical set of units to study the entirety of the lake basin condition, the watershed hydrologic units (HUCs) that comprise the entire drainage area of the land into the lakes themselves. As we outlined in previous chapters, the basic ecological theories and concepts that underlie landscape ecology are somewhat different from the fundamental elements of traditional ecology, in that the discipline of landscape ecology takes into account the spatial arrangements of the components or elements that make up the environment (e.g., the watersheds, patches of forest, and urban areas in the Great Lakes Basin). As we consider this larger landscape for analysis, bounded by the union of the contributing hydrologic watersheds of the Great Lakes, realize that some relationships between ecological patterns and processes can change, depending solely upon the scale at which the observations occur, and in the case of the Great Lakes Basin, there are multiple scales of importance to account for, from fine scale to broad scale, as you will observe among the Great Lakes metrics utilized in Chapter 4. As we understand from the prior chapter discussion, landscape ecology parameters include both the analyses of elements of nature as well as the presence and activities of humans as integral combined components of the environment (Jones et al. 1997) and, in the case of the development of landscape metrics for the Great Lakes, include important measurements that incorporate the long and complex history of human habitation developments, the development of industrial sectors, and the heavily agricultural activities that dominate much of the Great Lakes Basin.

Although the Great Lakes Basin is a relatively large portion of the Earth's surface, it is important to realize that a landscape is not solely defined by its size, but rather by an interacting mosaic of elements (e.g., ecosystem types), and in the Great Lakes, this is very relevant in that it encompasses many of the ecosystem types found throughout temperate regions of the planet with some of those ecosystems predominating the ecological functions of the entirety of the Great Lakes Basin, specifically the wetlands and upland forests of the region. Because of the importance of wetlands, forests, and human land use in the Great Lakes Basin, in particular their unique influence on ecological functions and services, much of the focus of the metrics is on wetlands, forests, as well as the human-inhabited areas within the landscape. Note how both remote sensing and landscape ecology provides the ideal theoretical framework for analyzing the complex and interacting spatial patterns of wetlands, forests, and human-inhabited areas throughout this and the following chapters and you will see that the landscape approach enables the articulation of ecological conditions and ecological risks for an otherwise difficult-to-describe and vast area of the world. Also, note that, at the broadest scale, the incorporation of multiple metrics allows for a better understanding of the entirety of the Great Lakes as a single ecosystem, that is, the Great Lakes Ecosystem.

As we feature the details of the Great Lakes project, notice the emphasis on the importance of remote sensing–based landscape ecology approach as one of the key techniques for developing indicators of overall ecosystem condition, as we strive to solve the real-world problems of protecting and restoring these areas as has been made a priority by the GLRI, while assisting in the formulation of solutions that are beneficial to the public who live in the Great Lakes region. The approach taken in the Great Lakes region is also quite applicable to other large projects in other landscapes around the world, and many of the essentials from this project are built upon in the subsequent projects outlined in Chapters 4 and 5. For a variety of reasons in the Great Lakes region, prior to this project, natural resource managers, regulatory agencies, and research communities have found it a challenge to complete a full basin-wide assessment of the Great Lakes, especially because one of the important ecosystem types, wetlands, have been extremely difficult to characterize in terms of their ecological condition across such a vast area (GLCWC 2004a). Some of the difficulty in accomplishing this characterization is related to the scarcity of complete data sets and other useful information regarding wetlands and the landscapes they are embedded within, across the entire region. Such a characterization of ecological condition even in limited areas of the watershed is difficult to accomplish, leading to an incomplete coverage of available and relevant data to accomplish (less than half of) the ecological measures that were identified in the 1998 State of the Lakes Ecosystem Conference (SOLEC 2000, 2009). This fundamental shortage of comprehensive information about wetlands is similar in scope to the shortage of full ecosystem/watershed data sets, which is at the heart of why there is no comprehensive long-term strategy for assessing the condition of the full Great Lakes. The approach we use in this project uses a traditional landscape ecology approach (Forman and Godron 1986; Turner et al. 2001; Brown et al. 2004) to address this lack of information to some degree, and investigate ecological conditions within the landscape of the Great Lakes Basin, specifically by utilizing selected landscape metrics as potential indicators of ecological conditions across the large area of this diverse landscape. Satellite and airborne remote sensing platforms, and the improving and available geospatial data products of the past decade or two, help to improve the circumstances of unavailable data and the theoretical basis for a landscape ecology approach has allowed the characterization of landscape conditions and processes.

Researchers had previously used a landscape approach to conduct simple regional assessments of environmental conditions, using some of the assessment results to further their goals of determining the interaction between landscape patterns and the flow of water, energy, nutrients, and biota in multiple ecosystems (USEPA 2001a). Information about the influence of ecosystem patch size, shape, and connectivity or the impact of human-built areas on ecosystems (Figure 3.1) has also been more fully developed and used to develop indicators of the condition of other things on the ground, and the Great Lakes Basin provides a perfect opportunity to utilize these metrics to

FIGURE 3.1
The size, configuration, and connectivity of natural and built areas within the landscape matrix provide important information about the condition of these entities themselves or as an integrated system; for example, those land cover and land use types immediately adjacent to or near forested areas, coasts, wetlands, and urban/built areas.

hypothesize the ecological condition, where human activities are elevated. Valid indicators can reveal dominant ecological changes with the most efficient use of resources, but cannot be used to determine the ecological condition at very fine scales, for example, a specific 12-hectare wetland reserve. Thus, landscape metrics of ecological condition can provide a good initial basis for assessments of ecological condition across a vast area like the Great Lakes and these metrics can be substantiated using scientific methodologies at finer scales and with field data. In the Great Lakes project, the intent of the initial approaches described are to produce a much-needed broad-scale "synoptic" view of the entire basin with regard to a variety of traditional metrics, which allows for developing further such hypotheses that lead to finer-scale and field-based verification work.

3.2 Metrics and Indicators

The terminology used in this project, and for the other landscape ecology projects in this book, is very specific and is worth using consistently to avoid confusion and ambiguity in meaning. It is also worth reviewing briefly prior to moving forward with an interpretation of the broad-scale ecological analyses in the Great Lakes. Although we've discussed these terms quite a bit up to this point, and you may have some familiarity with them, some additional attention to these terms is taken here as an opportunity to specifically highlight each of them and to be very intentional about explaining their use in terms of the Great Lakes project.

Standard measurements of ecological resources comprise ecological *metrics*. When measured at a relatively broad *landscape scale* (Forman 1995), *ecological metrics* (such as the percent cover of forested land cover in a location) can be described as a *landscape metric*, that is, a measurement that describes the condition of an ecosystem's critical components (O'Neill et al. 1992). For the sake of this project discussion, we will use the State of the Great Lakes Ecosystem Conference (SOLEC 2000, 2009) definition of an *indicator*:

> A parameter or value that reflects the condition of an environmental (or human health) component, usually with a significance that extends beyond the measurement or value itself. Indicators provide the means to assess progress toward an objective.

Indicators can be utilized as pieces of evidence, or clues, which provide us information about the condition of some environmental feature of interest (GLNPO 1999; Plexida 2014; Semadeni-Davies et al. 2014). Indicators have significance far beyond the actual values of the attribute measured because they describe the condition of some other function or value, by virtue of their relationship to the other condition. An indicator is typically a value calculated by statistically combining and summarizing relevant data. A simple example of the indicator concept is an economist's use of interest rates and unemployment to assess the status of economies, that is, economic indicators. Economists make seasonal adjustments for these indicators with a model, and most look at several indicators together instead of just one at a time (Jones et al. 1997). Similar to economic indicators, *landscape indicators* provide pieces of information that may tell us something about the broader condition of the ecosystems within a landscape. An *ecological indicator* is defined as a sample measurement, typically obtained by collecting samples in the field of an ecological resource (Bromberg 1990; Hunsaker and Carpenter 1990; Næsset 2002). For example, collecting plant material in a meadow for further measurements with a spectrometer may provide information about the amount of trace metals in the soil of the meadow, indicated by the concentration of those trace metals in the leaf of the plant.

Landscape metrics can therefore be used to characterize the environment at a broad scale, and they can also be used to develop field-verified landscape indicators (Jones et al. 1997), including indicators of habitat quality, ecosystem function, and the flow of energy and materials across the vast Great Lakes Basin. Among the best ecosystem candidates for developing these landscape indicators in the Great Lakes are wetlands and forests, primarily because they are relatively abundant along *gradients of condition* that range from least impacted to potentially impaired. These gradients of condition with a variety of wetlands and forests along them allows for determining gradients of ecosystem condition. Some of the relationships that can be analyzed, by correlating gradients of landscape condition with the range of ecosystem condition, can be hypothesized using fundamental ecological theory, including patch dynamics (such as the size, shape, and orientation of wetlands or forests) or processes (such as percent cover of vegetation, net primary productivity, or carbon content). For landscape indicators to be verified, a field sampling of the actual conditions within the focal ecosystems (wetlands and forests in this example) would need to occur to ensure that they are suitable as true landscape indicators of ecological condition in the Great Lakes region.

Of key importance is the recognition of which scale a metric (ecological metric or landscape metric) is being applied so that the results of such analyses can be viewed in the context of actual conditions on the ground, such as in the Great Lakes project's ecosystem foci, in forests and wetlands. Many land cover gradients are subtle; however, the data used for certain metrics may not be appropriate for capturing such subtleties, resulting in the inability to measure the true land cover gradient on the ground. For example, even though plants may be good indicators of soil trace metal concentrations in wetlands, field collection of 20 plant samples throughout a wetland (analyzed in the laboratory to determine the concentration of trace metals in the leaves of each) may be inadequate to determine the concentration gradient(s) of trace metals across an entire wetland. This is a similar issue to the problem that occurs at broader landscape scales with remote sensing and GIS data; if land cover data is provided at a 1-kilometer pixel size, that resolution of GIS data may be too coarse to measure the true gradients on the ground (e.g., small wetlands or small forest patches may be missed). Thus, two important guidelines to effectively avoid this pitfall, and for more effectively using landscape ecology metrics and indicators at relatively broad scales in the Great Lakes Basin are:

1. Select the most appropriate remote sensing or other data for addressing the ecological process or "endpoint" of interest.
2. Select the geospatial model(s) that is (are) most appropriate for detecting or describing spatial or temporal change in the landscape.

Selected landscape ecology endpoints and models can also be adjusted or modified to help interpret measurements, and to better understand overall ecological conditions (Jones et al. 1997; Pastorok et al. 2016) as improved data and understanding of ecological processes emerge and develop.

Over time, landscape metric and indicator values can also provide information on the trends in the condition of the ecosystem components. The information about trends in the condition of the Great Lakes Ecosystem helps to determine:

- If it is necessary to intervene in the landscape processes of the Great Lakes
- If so, which intervention in the region will yield the best results, and at what extent(s) within the larger Great Lakes landscape
- How successful interventions will be implemented, once landscape level actions have been taken within the region

3.3 Landscape Scale and Gradients

From the previous discussion, it is apparent that environmental gradients are important to use in the Great Lakes for a better understanding of the relative condition of ecosystems among watersheds of the Basin, keying in on least impacted to impaired locations, for example, and the scale of analysis determines what measure we might find relevant. Scale is generally defined by the extent of information, and the grain of information being analyzed. The extent of information is the spatial domain, or the size of the area studied, for which data are available (McGarigal 2002). Grain of information refers to the minimum resolution or size of the observation units, often identified as digital picture elements (i.e., "pixels") or patches that are made up of aggregations of pixels. When these patches are determined, the pattern detected in any portion of the landscape is a function of the scale of observation and mapping. As the most fundamental sampling unit used in remote sensing and raster data, the pixel is generally a square element, having four sides or edges. A pixel can also be an equilateral triangle, a hexagon, or any other equilateral tessellation. A contiguous cluster (using the cardinal rule) of homogeneous pixels is referred to as a *patch*. For example, a patch can be a cluster of pixels that define the area of a lake, a tract of forest, or any other group homogeneous area with respect to a specific classification (e.g., water, forest, or wetland). A mosaic of various patches across a given area constitutes a landscape and is sometimes referred to as a *landscape matrix*, within which patches are embedded.

There are a number of important terms that describe the characteristics of patterns found within the landscape matrix that are particularly important in the Great Lakes project. These terms include diversity, dominance, contagion, fragmentation, and patch shape (also see the initial background on these terms in Chapter 2), and have very practical application in the Great Lakes. Each of these terms has a variety of definitions in the landscape ecology, geography, and remote sensing approaches used in the Great Lakes project. The diversity of a landscape can be described in several ways, such as the total number of land cover types. This specific definition of diversity is often referred to as landscape richness (Forman and Godron 1986; Forman 1995). Dominance is defined as the degree to which one or a few land cover types predominate within a given landscape in terms of proportion (Forman and Godron 1986) and can be calculated in a number of different ways depending upon needs and applications.

In a practical sense, as applied in the Great Lakes, the next two terms, contagion and fragmentation, are opposites of one another. *Contagion* is the tendency of land cover types to be clustered or clumped into fewer larger patches of that land cover type (Wickham et al. 1996). On the other hand, *fragmentation* is the tendency of land cover types to be broken up into greater numbers of smaller patches (Forman 1995) of that land cover type. A landscape with high contagion would be one with low fragmentation, and a landscape with high fragmentation would be one with low contagion. Thus, even though contagion is not defined as a measure of fragmentation, it is directly related to fragmentation. It is important to understand that both terms, contagion and fragmentation, are dependent upon the spatial resolution of analyses. For example, a 30-m spatial resolution landscape representation with a particular patch configuration might be considered fragmented. Whereas the same patch configuration within a landscape analyzed at a 1-km spatial resolution may be considered to have a high contagion, and thus very low fragmentation. These spatially dependent results may also be dependent upon the ecological processes being analyzed within the landscape matrix. For example, the landscape may be considered to have high contagion and low fragmentation for a large predatory bird that has a broad feeding range, while the same landscape may be considered fragmented and have low contagion for a relatively smaller bird, such as an avian flycatcher.

Patch shape (i.e., complexity), refers to the relationship between the perimeter of a patch and the area of the patch. Again, these concepts were reviewed in depth in Chapter 2 and in this chapter we introduce the concepts in terms of the practical applications to projects, such as in the Great Lakes. In general, complex patches are those that have greater scaled perimeter-to-area ratios, while simple patches are those that have lower scaled perimeter-to-area ratios. Once again, this term is relative to both spatial resolution and ecological process. A key element you will read about throughout this book is the need to better understand the relationships between remote sensing–based landscape metrics and the specific ecological processes represented by those landscape metrics, and, importantly, the constraints of determining these relationships.

Landscape ecologists carefully consider the optimal scale of the information they will use in their analyses and the gradients used for analysis of land cover data or other biophysical data, depending upon the questions that are being addressed by a project, as in the Great Lakes. In order to understand risks to ecological resources and humans, it is also important to analyze the spatial patterns of environmental conditions at a variety of relevant scales, for example, ranging from a single plot in a forest patch to a large region of the landscape, such that the scales are relevant to the questions posed. Scientists may also select metrics and indicators that reflect environmental conditions on a variety of scales in both space and time and then determine how closely trends track with each other to determine if trends are scale dependent, and to what degree. This is a technique utilized to good effect in the Ozarks Mountains case study (Chapter 5), where water quality is tracked similarly at a variety of scales, resulting in an indication that riparian vegetation condition is correlated with water chemistry and microbiology in streams downstream or downslope from those vegetated areas.

A landscape ecological project requires a definition of the scale of the input data (e.g., 30-m pixel size for land cover), and requires the user to understand what scale is appropriate for their particular application (e.g., minimum path size that is relevant to species requirements). It is also an important responsibility of the user (and for project teams to guide them) to exercise caution when attempting to make decisions at, or among, different scales of landscape ecological outputs. For example, in the Great Lakes, wetlands that are smaller than 900 m^2 (i.e., the minimum pixel resolution of the National Land Cover Database's base imagery, i.e., Landsat) are likely too small to have been detected in the land cover classification process, and even slightly larger wetlands may be missed in the classification process because of factors related to the physics of the satellite sensor system used in the production of the land cover data and patch shape metrics. Therefore, broad-scale monitoring of such small wetland areas may be irrelevant by directly observed landscape metrics using the NLCD (Lopez et al. 2003, 2006a). Note that, generally, when it comes to the resolution of remote sensing imagery, "fine resolution" usually refers to relatively fine-grain pixel resolution (e.g., centimeters to 5-m pixel size, currently) and "coarse resolution" refers to relatively large-grain resolution (e.g., greater than 5-m pixel size, currently).

Given all of the factors involved, and especially across a vast and diverse landscape such as the Great Lakes Basin, it is important to select gradients (i.e., changes over space or time) of condition(s) that offer sufficient variability, and a sufficient number of field-sampling sites to compare among reporting units (Green 1979; Karr and Chu 1997; Lopez et al. 2002), particularly if those ecological metrics are to be used to develop landscape indicators. Landscape (e.g., land cover) gradients may be useful for the development of landscape indicators because the statistical relationships between landscape metrics and ecological metrics can give clues about how two (or more) elements of the landscape may interact, such as the relationship between agriculture in a watershed and the concentration of phosphorus in wetlands, streams, and rivers. In addition, the use of previously observed in situ correlations between biophysical measurements may help to guide the analyses of relevant parameters that may be good indicators of ecological vulnerability at moderate to coarse scales. These approaches are all utilized in the Great Lakes to the best effect possible, by specifically analyzing a large area of the landscape, optimally utilizing a multitude of gradients of condition across all of the watersheds of the Great Lakes Basin, and utilizing the finest resolution data available at that vast extent (i.e., 30-m pixel size; see Figure 3.2).

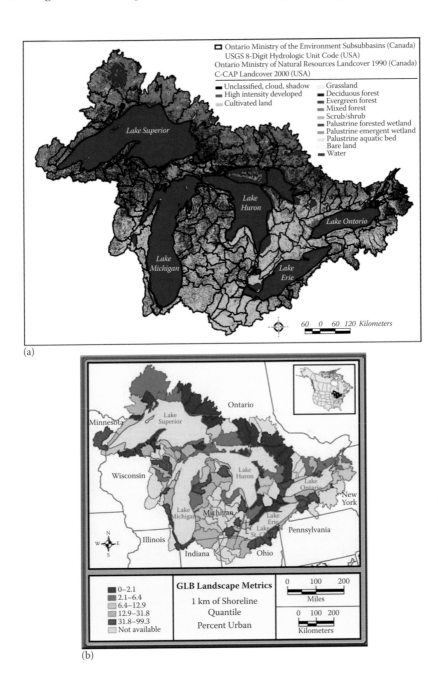

FIGURE 3.2

An example of how (a) land cover derived from satellite remote sensing data has been used to produce (b) metric maps in the Great Lakes Basin. These metric maps can then be used to develop indicators of ecosystem condition. This specific example shows the use of the "High Intensity Developed" data classification to calculate and map "Percent Urban" among each of the Great Lakes Basin's watersheds, within 1 km of the shoreline.

It is important to note that prior to the advent of contemporary remote sensing, data processing capabilities, and GIS technologies, it was prohibitively expensive and time-consuming to calculate metrics of landscape composition and pattern at multiple (spatial or temporal) scales throughout the entirety of a vast area of the landscape like the Great Lakes, and essentially an impossibility due to the complexities of processes involved. Without a full understanding of the spatial and temporal patterns of landscape composition and pattern (Figure 3.3), knowledge of the condition of wetland and forest ecosystems, and the vulnerability of these natural resources to loss and degradation, is extremely limited. Now, with the use of landscape metrics, correlations with ecological metrics collected in the field, leveraged with the use of statistical tools, can be used to determine the association between the broad-scale data (i.e., landscape metrics) and the fine-scale conditions on the ground (i.e., ecological metric). A determination of correlations between the broad-scale (e.g., Riitters et al. 1995; Jones et al. 2000, 2001), moderate-scale (e.g., van der Valk and Davis 1980; Roth et al. 1996;

FIGURE 3.3

An oblique aerial view of a coastal wetland complex in the Great Lakes Basin. Coastal wetland complexes like this are important landscape elements and their ecological functions provide many ecosystem services, such as water quality improvement and storm/flood protection. Ecosystems like coastal wetlands can be described as patches, vegetation associations, and with metrics that describe the size, topographic position, interspersion, orientation, and relative proximity to other elements in the landscape.

Nagasaka and Nakamura 1999; Lopez et al. 2002), and fine-scale (e.g., Peterjohn and Corel 1984; Murkin and Kale 1986; Ehrenfeld and Schneider 1991; Willis and Mitsch 1995; McIntyre and Wines 1999a; Luoto 2000) measures has not been completely explored, however, is among the critical research issues being addressed by landscape ecologists in close coordination with remote sensing scientists, as was the case in the Great Lakes project, and as happens in other areas where decision making at a variety of scales is needed (e.g., for municipality, county, state, and federal land use and planning ordinances, laws, and regulations). One of the fundamental bases of the work in the Great Lakes is the existing list of potential and operational indicators of condition for wetlands (at several scales) in the Basin (SOLEC 2000, 2009).

3.4 Landscape Models

Because the Great Lakes Basin (i.e., the landscape area that delineates the land's drainage area and lake basin itself) is very complex, an initial focus on the relevant biophysical characteristics (i.e., excluding few of the possibly relevant biophysical characteristics) is an important first step toward developing landscape indicators. GIS is a key tool that can be used to focus on relevant features of the landscape. For example, a GIS-derived landscape metric, such as percentage of cropland area among watersheds, can be correlated with water quality parameters either conceptually or statistically if the data are available at a location that is known to be the outlet of a watershed, and a geostatistical model can then be developed. The relationships might be analyzed as a causal (predictive) relationship, perhaps using a regression model with watershed condition as the independent variable(s) and water quality parameter(s) as the dependent variable(s). The causal relationships of these variables might be based on a priori knowledge acquired as a result of previously published in situ studies of similar variables, and ecological theory as a whole. Broad-scale models such as those applied to the Great Lakes, founded on the ecological principals of in situ studies, may be limited by a lack of detailed information about small areas, but can serve as a preliminary tool to assess large areas that would otherwise be impractical to assess in the field, or where full coverage of detailed GIS data is absent. Response variables may also be tested (e.g., water quality measurements, habitat characteristics, or wetland functional characteristics), depending on the objectives of the user, and the ecological endpoint of interest. These relationships are explored more fully in the Ozark Mountains project and the Kansas River project in Chapter 5.

3.5 Understanding the Benefits and Challenges of Using Selected Landscape Metrics and Approaches

Because landscape ecology is the study of the effects of landscape patterns on ecological processes (Turner 1989), it can be frustrating for some to discover that there is not always perfect field-based data to complete the analyses of a particular region of interest. This unfortunate situation can be tempered by the understanding that methods are constantly being developed to leverage various remote sensing data and analysis techniques to quantify landscape patterns so that measurable links between observed patterns and ecological processes can be determined for a project, despite imperfect field-based data coverage. The Great Lakes project is a prime example of this situation, where there is a tradeoff between a broad view of the landscape, very valuable and new, but consequently a limited characterization of all aspects of the landscape due to limited consistency and abundance of field-based data across the vast area. The most common method for overcoming this situation is quantifying landscape patterns by capturing information of particular spatial patterns based upon available data and characterizing the patterns as a single value, for a reporting unit, such as the watersheds in the Great Lakes project. Such values are referred to as a landscape metric (the measure itself) or a landscape index, a selection of multiple metrics, often mathematically or statistically combined. Multiple examples are provided in subsequent chapters of this book, with some innovative geospatial statistical methods, such as the use of partial least squares analyses, used in the Ozark Mountains project (Chapter 5).

Landscape metrics and landscape indicators may be used to assess progress toward one or more objectives, and these have been determined by the SOLEC (2000, 2009) and others (Smith et al. 2015), in the Great Lakes to determine ecosystem condition. Thus, the selection and use of metrics and indicators in the Great Lakes is guided by the purpose for which the information will be used in the region, and these reasons have been discussed in depth and among many in the professional fields within the entire Great Lakes region for years, whether research-oriented or policy-oriented. Depending upon the use, the relative importance of quality, cost, and completeness of the coverage of the metric or indicator has also been a key decision node for pursuing the development and use of a particular metric or indicator, with cost as a key factor for implementing for a particular project.

Naturally, there is a crossover between the goals of the pure research and policy motivators for pursuing particular metrics and indicators, such as in the Great Lakes. The crossover between landscape ecology science and policy is a result of the common primary goals of each perspective, which are essentially both focused on identifying key indicators of ecological condition that can serve as sentinels of important change. Despite this common primary goal, the differences between the scientific and policy uses of landscape ecology can be profound, and are generally a result of differing secondary goals. The common primary goals of the research and policy communities are what are key to successful projects, as in the Great Lakes, and are summarized below (note: all topics are being addressed to some extent by both science and policy groups, although not always as a primary goal):

1. Assess changes in the condition of the ecosystems and the progress toward achieving management goals for its sustainable well-being.

2. Improve understanding of how human actions affect the ecosystems and determine the types of programs, policies, or regulations needed to address the environmental impacts.

3. Gain a clearer understanding of existing and emerging environmental problems and their solutions.

4. Provide information that assists the public and stakeholders in participating in informed decision making.

5. Provide information that will help managers better assess the success of current programs, and provide a rationale for future ones.

6. Provide information that will help set priorities for research, data collection, monitoring, and cleanup or restoration programs.

3.6 Regulatory Support Uses in the Great Lakes (Policy Goals)

As environmental regulations were initially being developed in the United States, there was a focus on the established measures of environmental quality, such as those for drinking water and air quality. These measures reflected a traditional view of the environment and the potential for multiple factors that may contribute to environmental degradation (Jones et al. 1997). Research that was supported by regulatory agencies addressed the need to make policy recommendations to decision makers, but did not fully address the scientific (i.e., ecological research) community's goal of increasing our understanding of the interrelationships between abiotic and biotic parameters (Zandbergen and Petersen 1995).

Thus, landscape indicators were initially developed as lists of physical and chemical measures to monitor improvements in water quality. Biological responses resulting from changes were not considered. Requirements for environmental impact statements led to development of procedures to evaluate habitat as the basis for environmental assessment. As government policies endeavored to protect both human health and the environment from the by-products of an industrial economy, scientific research required a different approach to support these policies. Awareness of the scope of environmental problems increased and toxic substances became a concern. A variety of tissue, cellular, and subcellular indicators were developed as diagnostic screening tools or biological markers, to evaluate the physiological condition of an organism and to detect exposure to contaminants. The need to develop management strategies to address interactions within ecosystems and the impacts of human activities at a broader scale, upon those natural systems, became another stimulus for the development of indicators.

The development and use of indicators that meet all of these needs in the Great Lakes and elsewhere tend to be a learning process for both the scientists who develop them and for the policy makers who use them. Scientific knowledge itself is the outcome of a consensus-building process among scientists from different disciplines who require easily interpretable descriptions of ecological condition (Zandbergen and Petersen 1995). Developing landscape indicators in the Great Lakes involves the collection and management of supporting data, the identification and use of selection criteria, the evaluation of indicators for their efficacy, and accounting for the influence of scale on the final product. Landscape indicators are also an important input to Decision Support Systems for the Great Lakes, which can be utilized by policy makers and environmental professionals who require the most up-to-date and accurate information for determining effective strategies for ecological monitoring, assessment, restoration, characterization, risk assessment, and management.

3.7 Ecological Research Uses in the Great Lakes (Science Goals)

The current ecosystem concept and approach to studying ecological interrelationships was conceived as a multidisciplinary, problem-solving concept with the goals of restoring, rehabilitating, enhancing, and maintaining the integrity of particular ecosystems. The answers to scientific questions posed by the utilization of remote sensing data, and a broader-scale viewpoint within and among ecosystems, have created a new list of scientific questions that focus on how these ecosystems interact with the surrounding biophysical environment, and thus have spurred a new area of investigation into the pattern of land cover, and the implications of that land cover pattern on ecosystems in the landscape (e.g., a forest that is embedded in a larger landscape of agricultural cropland). Not all of these relationships have been fully field tested across vast areas, yet many landscape indicators have been conceptually proposed and have been tested, that is, developed from theoretical ecology (USEPA 2001a; Lopez et al. 2006a) in the Great Lakes. Relatively few results are available that show comparisons of landscape metrics or metric performance at different scales (Cushma and McGarigal 2004), but some of these relationships in the Great Lakes, and subsequent work in Chapters 4 and 5 help to move this work forward, and several new patterns have been developed to advance this field of work in the Great Lakes Basin as well, especially as it pertains to forest and wetland ecosystems. To this end, the work of a subgroup of participants in the Great Lakes Commission's Great Lakes Coastal Wetlands Consortium (GLCWC 2004a) has taken some initial strides forward to compile those biophysical measurements that directly relate to ecological endpoints within ecosystems and watersheds of the Great Lakes (which apply to more than solely coastal wetlands), as follows:

1. Amphibian community condition
2. Areal extent of wetlands by type
3. Bird community condition
4. Contaminant accumulation
5. Extent of upstream channelization
6. Fish community condition
7. Gain in restored wetland area by type
8. Habitat adjacent to wetlands
9. Human impact measures
10. Invertebrate community condition
11. Land use classes adjacent to wetlands
12. Land use classes in watersheds
13. Phosphorus and nitrogen levels
14. Plant community condition
15. Proximity to navigable channels
16. Proximity to recreational boating activity
17. Sediment flow and availability
18. Water level

The 18 specific GLCWC metrics are the basis of the Great Lakes landscape metric work outlined in Chapter 4, which takes these conceptual elements and translates them to the entire landscape of the Great Lakes, for the first time; the metric work of the Great Lakes Commission and the work in the Great Lakes case study in Chapter 4 has also been used as the basis for additional work in the Great Lakes that delves even further into the use of remote sensing to explore ecological condition, such as invasive plant species (Bourgeau-Chavez et al. 2015; Marcaccio and Chow-Fraser 2014).

TABLE 3.1

Floral and Faunal Indicators and Methods for Obtaining These Measurements
for Initial Assessment in the Great Lakes Ecosystem

Indicator (SOLEC ID)	Measurement Description	Method Summary
Invertebrate community health (4501)	Diversity indices, adult caddisfly presence/absence and diversity.	Sweep nets, activity traps, backlighting, Hester–Dendy samplers. Need standardized processing. Need standardized habitat sampling. Repeat visits.
Fish community health and DELTs (4502, 4503)	Several diversity and abundance (fish per meter) measures, incidence rate of DELTs (deformities, eroded fins, lesions, and tumors).	Electroshocking along transects, fyke nets.
Amphibian diversity (4504)	Many possible population, diversity, and abundance measures. Compare with extensive measures. Species presence, abundance, and diversity.	From most intensive to most extensive—complete counts, capture-recapture, larvae sampling, drift fences or pitfall traps, funnel trapping, visual encounter surveys, Marsh Monitoring Program, and audio surveys.
Bird diversity and abundance (4507)	Intensive—many population, diversity, and abundance measures. Compare with extensive measures—species presence, abundance, and diversity.	Intensive—territory mapping, strip censuses, nest counts, site inventories. Extensive—MMP survey.
Plant community health (4513)	*From air photos*: % dominant vegetation types, % invasive types; *from floristic survey*: % wetland obligate species, % native taxa, floristic indexes; *from quantitative sampling*: % cover of invasives in dominant emergent, % floating/ submersed cover of turbidity tolerant taxa, rate of change in invasive taxa.	Air photo compilation and interpretation, floristic survey, and quantitative sampling.
Contaminants (4506)	Contaminant levels or physical anomalies. *Further work is needed to develop this indicator.*	External survey of bullheads, DELTs, or other methods that provide useful biological contamination metrics.

The GLCWC also recommends the following six criteria (2004b) that should be applied when selecting landscape indicators that are applicable to the Great Lakes:

1. Cost and level of effort to implement basin wide
2. Measurability with existing technologies, programs, and data

3. Basin-wide applicability or sampling by wetland type
4. Availability of complementary existing research or data
5. Indicator sensitivity to wetland condition changes
6. Ability to set endpoint or attainment levels

Indicator summaries in this chapter (Tables 3.1 and 3.2) are useful in the initial conceptualization stages of developing landscape ecological indicators of any project, and indeed are the basis of the Great Lakes project, and these summaries ensure that geospatial data will adequately address ecological endpoints. The indicators included for the Great Lakes are comprehensive and include both flora and fauna (Table 3.1) and physical measurements (Table 3.2). Field verification is necessary for validating these metrics, as envisioned by the GLCWC, for they are based on ground-level phenomena. Field methodologies can also be modified to directly address the ecological

TABLE 3.2

Physical Characteristics and Methods for Obtaining These Measurements for Initial Assessment in the Great Lakes Ecosystem

Indicator (SOLEC ID, as available)	Measurement Description	Method Summary
Water levels (4861)	Lake levels, wetland water levels, in-/outflows	Data obtained from lake gauges.
Sediment flow (4516)	Suspended sediment unit area yield (tons/km^2 of upstream watershed)	Metric should be estimated from gauging stations upstream of wetland. Sediment core or turbidity measures.
Sediment available for coastal nourishment (8142)	Sediment budget, net accumulation/loss	Metrics measured from streamflow and sediment gauging stations at mouths of major tributaries. Alternatives—geomorphic surveys of barrier bars/islands, air photo interpretation.
Storms and Ice	Possible metrics include wetland form factor, succession lag times, storm erosion of shore buffers; ice cover duration, ice thickness, ice jams.	Methods vary by metric.
Phosphorus and total nitrates (4860)	Total phosphorus and nitrates concentrations from May to July for correlation with other metrics. *Further work is needed to develop this indicator.*	Metric calculated from concentration and flow measures from gauging stations.

endpoints of the specific sensitivity of the landscape metrics/indicators, as needed. The approach taken in the Great Lakes case study (Chapter 4) was, therefore, to take these metrics further by utilizing remote sensing data, and drawing some general conclusions about watershed condition for several of these proposed ecological indicators, in the coming chapter, based on the landscape metrics.

3.8 Inferring Ecological Condition

A landscape ecological assessment is not merely providing some measurements of land cover, but is also a determination of the condition of an entire region, such as the Great Lakes Basin, and ideally includes reference to both specific ecosystems and their surroundings. This should involve the general summary of conditions, a determination of suitability of habitat for specific plants or animals, a determination of the health risks for humans (e.g., water quality, erosion potential, or other impacts), or the determination of vulnerability for specific plants and animals (i.e., the ecological vulnerability). Such determinations ideally require a basis of comparison (a benchmark), which might be based on the least-disturbed condition that is known or is desired for the ecosystem. Luckily, with the long-standing work of groups in a region, such as the GLCWC in the Great Lakes, there is a wealth of information to develop indicators from both qualitative and quantitative inference. What is meant by qualitative inference is the understanding that, for example, larger patches of forest provide more foraging, resting, or mating habitat for certain bird species as a general rule than smaller patches of forest. These inferences do not involve a statistical correlation, although that may be possible with sufficient field data on bird activity, but rather rely on ecological theory and general qualitative understandings, such as the understanding and ecological theory behind the concept that larger areas of identified appropriate land cover types lead to greater habitat availability for organisms that frequent those same land cover types, and the converse. There are several qualitative inferences outlined in Chapter 4 for the Great Lakes, contrasted with the quantitative inferences outlined in Chapter 4 for the Ozark Mountains. This contrast in the two different situations and approaches driven by those circumstances of data availability, as well as the availability of ecological theory and practical knowledge, are instructive in terms of understanding the limitations that exist for landscapes with relatively minimal field-based data, such as in the Great Lakes.

In general, the condition of the assessment area, or a portion therein, can be compared to the benchmark condition (sometimes called a "reference condition") to establish specific criteria against which proposals for change or modifications might be presented. Although past impairments may preclude restoration of any given ecosystem to natural conditions, the perceived natural condition must be understood in order to define the target condition and guide ecosystem improvement. That is, in the Ohio coastal area (for example) it is impossible to find wetland ecosystems that are currently in the condition that they might have been found in the mid-1700's (i.e., pristine), but this is not to say that natural wetland conditions that may be desirable, but are not currently present in the landscape, cannot be used as a reference condition to compare to the current conditions found throughout the landscape. Thus, reference conditions are frequently thought of as the "least impacted" area or conditions for a specific geographic region. Understanding specific impairments to physical, chemical, and biological conditions is a precursor to determining appropriate improvements. Movement from any current condition toward a reference condition would be considered an improvement; movement away from the reference condition might be considered harm or degradation.

3.8.1 Inferring Habitat Suitability and Vulnerability

Habitat information about plants and wildlife species is frequently represented by scattered data sets collected during different seasons and years, and from different sites throughout the range of a species. A GIS-based model of habitat suitability (i.e., based on the physiological and sociobiological requirements of a species or taxa) and habitat vulnerability (i.e., potential for degradation) can present an otherwise complex database in a formal, logical, and simplified manner. Habitat models are a formalized synthesis of biological and habitat information and include many assumptions about the organization of the model components. Thus, such models should be regarded as hypotheses of species-habitat relationships, and not as a statement of proven cause and effect relationships, unless the metric model has been thoroughly tested using a valid statistical design. Habitat models may have merit in planning wildlife habitat research studies about a species, as well as providing an estimate of the relative suitability of habitat for that species (Rogers and Allen 1987; Lopez et al. 2003, 2013).

Data needs for developing and using a GIS habitat suitability or vulnerability model may include bathymetric and topographic maps, aerial photographs, categorized satellite imagery, surface water area, streamflow, river stage, species-specific habitat requirements, historical precipitation, air or water temperature data, and the potential for future precipitation and temperature. Vulnerability can be assessed by comparing habitat quality, availability, or distribution with historical and projected future conditions.

3.8.2 Inferring Water Quality and Hydrologic Impairment

Certain ecosystems and land areas require special attention because of their unique position in the landscape, which makes them a particularly important ecosystem for intercepting, transforming, and accumulating chemical constituents that flow from upland areas to the open water areas of the lakes (e.g., wetlands, urban areas, upland forests, and agricultural lands). As runoff in the upland areas of Great Lakes Watershed passes through or is stored, natural land cover types often transform and retain nutrients (e.g., nitrogen and phosphorus), some pollutants (e.g., pesticides or components of road runoff), and reduce the amount of sediment that might otherwise be transported beyond the coastal areas to open water areas.

Different natural land cover types have unique roles in hydrologic change within the landscape, and their presence may ameliorate the erosional forces associated with hydrologic flows in upland areas. Changes in the hydro-period of the lakes and wetlands (that is, altered patterns of water levels and flows in and out of the wetlands), for example, can quickly lead to a change in the vegetated communities of wetlands, which can, in turn, change the habitat structure of wetlands and potentially alter the flow of material in and out of the wetlands, and further downstream (see details of this phenomenon in the Kansas River Watershed in Chapter 5). Thus, the geomorphology or the hydrodynamics of wetlands, and the entire landscape for that matter, can be used to infer ecological conditions in both the wetlands, and in downstream/slope areas of the landscape.

3.9 Utilizing Complementary Research for a Successful Project

The Great Lakes Consortium, USEPA, and the members of the SOLEC originally served several interests in the investment of additional resources for the assessment, restoration, and understanding of the processes of wetland ecosystems of the Great Lakes, which has now come to fruition with the advent of the GLRI legislation, initiatives, and work. As part of their foundational efforts, these groups fostered and sponsored complementary research for years prior to the GLRI that involved the collection and processing of a tremendous amount of remote sensing and GIS data, as well as substantial amounts of field-based information, including those data discussed and utilized in Chapter 4 outputs for the Great Lakes Ecosystem. Each of these groups conducted and supported detailed field studies at a variety of locations throughout the Great Lakes Basin as a foundational activity, mostly at specific scientific field sites associated with universities and agencies in the region. Each of the groups also compiled much of these data at the end of each project so that cross-site comparisons could be made, and the assessment of the Great Lakes Basin could proceed efficiently in the event that there was a larger comprehensive effort that arose in the future, and indeed it did, in the form of the GLRI. A basin-wide examination of wetland ecosystems using remote sensing techniques was recognized as one of several common goals of these groups. Thus, the integration of sampling and analytical protocols and benchmarks for implementing an effective binational and basin-wide monitoring program, which was capable of tracking and assessing the existing status and projected integrity of Great Lakes wetland ecosystems, was considered to be a high priority for the consortium and associated collaboratives. Collaborations between the members of these groups were overlapping and offered many opportunities for complementary research that, individually, may have been less effective and less collaborative than the GLRI effort as a whole. Complementary work under the guidance of the consortium included the development of a monitoring database, implementing a monitoring plan, and coordinating implementation with consortium member organizations (GLCWC 2004a). Thus, a key step for this, and any, landscape ecology project team is always to assess present and past studies so they can avoid duplication of data acquisition and processing, and to build upon the work of prior studies. It is recommended that a team set goals for future collaborations, and always imagine that as a possibility when designing a landscape ecological project to think in terms of 10 to 50 years of collaborative work into the future, at a minimum.

3.10 Communicating Landscape Scale Projects Effectively

A landscape ecological analysis can take many different forms, but invariably it needs to address very specific needs and audiences. Thus, a cost-effective technique for conveying landscape ecological assessment results is desirable and can be achieved by:

1. Selecting the minimum necessary metrics to address the basic questions of the research, regulatory goals, or decision-maker's information needs.
2. Determining the key "next steps" that might be taken if additional funding is forthcoming, and which metrics might be the preliminarily assessed during the initial analysis stages of the project.
3. Reviewing and synthesizing the ecological, remote sensing, geostatistical, and other theory bases that might be necessary to explain the assessment results.
4. Utilizing the appropriate and most effective means of portraying results, in written or digital formats.

The culmination of the landscape ecological assessment requires a decision about what the best format for conveying the results of the study is, considering the research and policy goals of the audience. The USEPA's Office of

Research and Development developed the "landscape atlas" concept (Jones et al. 1997; Lopez et al. 2003), which communicates the complex analyses of remote sensing, GIS, and field-based information to a variety of users. The breadth and complexity of landscape ecological information may hinder some audience members when they are in need of immediate answers to their issues or in need of immediate information to convey to local and regional stakeholders. A landscape atlas integrates broad-scale analysis results into a format that reduces the volume of data to a series of preselected maps, with standardized legends. The landscape atlas format can thus offer a series of informative maps that give the reader a general picture of the variety of ecological parameters across a common region, and pinpointing specific areas of interest, depending on the audience's needs. These maps can be in either written or digital formats, or, increasingly, in app formats in which users can interact with the data on handheld devices.

It is also possible to explain ecological endpoints at a broad scale with greater ease by grouping the maps by common geographies or metric/indicator types. The atlas format allows for looking at assessment results among scales, and based on different important topics that relate to ecological vulnerability, among different maps and tables, and it offers a way for the reader to compare metrics in a way that is most useful for particular needs (Lopez et al. 2003). An effective map format is often designed to give the reader an idea of the spatial distribution of ecological conditions relative to specific environmental values, at multiple scales, or during different historical periods.

Internet-based decision support tools, or perhaps memory stick/mini-drive or CD/DVD-based (or printed/plotted) equivalents for those working in areas of the world with limited or no Internet connectivity can help to incorporate the maps described above in a format that is readily accessible to those using a personal computer, depending on the availability of peripheral computer hardware or Internet connection speed/reliability. Note that approximately 60% of the world's population has no current Internet access (http://www.internetlivestats.com/internet-users, checked January 25, 2017). For those with Internet accessibility, the advantage of technologically appropriate and enhanced modes of data access continues to enable users to view larger and larger volumes of data, simulate analyses of the data for their individual purposes, and download the data associated with maps so that they can perform their own analyses. Because data sets for the entire Great Lakes Basin require a tremendous amount of storage space and a tremendous amount of processing capability, it is preferable to prepare all of the possible metric maps for an area, and then deliver them to the public in one of these digital or online formats. Again, it is very important to recognize that the majority of users and decision makers around the world have a need for formats that can serve communities with limited Internet (i.e., very slow or intermittent connections) and limited computer availability/skills, and a reiteration of the periodic need for a printed/plotted (laminated, often) version of key maps are valued by many, despite the fact that much of those working on a project are fully immersed within the modern electronic age. As data become more uniform across the Great Lakes Basin or as specialized collaborations develop, the use of Internet-based (e.g., Internet-based mapping applications) decision support tools are becoming more practical, in support of the GLRI and other efforts in the Great Lakes Basin. For additional information, visit the Great Lakes Restoration Initiative Internet website at https://www.glri.us (checked January 25, 2017).

Emblematic of the case studies to follow in Chapters 4 and 5, the presentation of remote sensing–based landscape ecology results, including a thoughtful and methodical outline of the steps taken to develop those landscape metrics and indicators that led to the successes of projects, like those in the Great Lakes project, requires several important steps, with a leader to guide them, including the following:

1. Gain knowledge of the data available for the region of interest
2. Use the knowledge of those who have experience in both the theoretical and the applied work that needs to be done
3. Consider what metric and indicators are possible given the available data and consider collecting new data, as needed and practical
4. Understand how to leverage scale and gradients of condition at the project site
5. Understand and articulate all policy goals and science goals prior to beginning the project
6. Do your best at inferring both qualitative and quantitative ecological condition
7. Utilize either a multidisciplinary approach, and if possible and cost effective, a transdisciplinary approach (Hirsch Hadorn et al. 2006) to ensure complementarity of approaches among all disciplines
8. Communicate thoroughly what you understand the projects to be, and convey that understanding to all of those who will be affected by the outcomes of the project
9. Ensure that the challenges that come along, such as they certainly did in the Great Lakes project, are approached as opportunities to grow the project team's understanding of the ecological relationships in the project area
10. Push the technological boundaries of remote sensing to a new level, and challenge the team leader to create a new template for future projects of similar types and needs

4

Applied Analyses of Broad-Scale Landscape Gradients of Condition

Two case studies are discussed in this chapter. One is relatively straight-forward in the use of data and metrics, but covers a truly vast area of the landscape—the Great Lakes Basin. This groundbreaking work provides analyses of entire geography even with the challenge of having to cover a very large area of the planet, with a land (a watershed) drainage area of 521,830 km^2 and a shoreline length of 17,017 km. An important outcome of the work in the Great Lakes is a summary look at ecological conditions, uti-lizing GIS data set partitioning to better understand ecosystem functions at a landscape level among and within several portions of the larger water-shed, specifically focusing on the hydrologic units of the watershed. As dis-cussed in the previous chapter, this broad-scale work helped to pave the way for much work that has proceeded within the context of the Great Lakes Restoration Initiative.

The second case study in this chapter utilizes landscape models to analyze impacts of global scale drivers of change, that is, sea level rise, on selected coastal areas (i.e., Coastal California and North Carolina), which may be widely applicable to vast coastal areas, worldwide. Both of the case stud-ies in this chapter provide a solid basis of the challenges and opportunities that are encountered at relatively broad scales, among vast areas and gradi-ents of conditions within a diverse landscape.

4.1 Case Study: The Laurentian Great Lakes Watershed

We progress in this and the next chapter through a series of increasing project complexity so as to provide key applied examples of the elements and approaches discussed in the previous chapters, specifically in this chapter by utilizing remote sensing to accomplish landscape ecologic goals across vast areas and among broad gradients of condition. The first case study provides the technical and analytical details of a project in the Great Lakes region, which serves as a demonstrative example of how the approaches discussed in the previous chapters can be used to meet the needs of broad-scale multiple-region analyses. This case study is focused on the ecology of the larger Great Lakes Ecosystem, emphasizing wetlands and forested natural areas.

The following project was designed to provide managers in the Great Lakes Basin with definitive answers to the following questions:

1. What basin-wide information is available for the development of landscape indicators?
2. How can remote sensing and GIS be used to develop landscape indicators of the ecological condition within the ecosystems of the Great Lakes Basin?
3. Is the available information sufficient to detect and analyze trends in landscape indicators for the Great Lakes Basin?

The answers to the above questions directly address the following concerns that are of most interest to the Great Lakes research and management communities:

1. What influences the cost of implementing basin-wide techniques for the use of landscape indicators of the ecosystem and landscape ecological condition?
2. What is the measurability of such techniques in the larger context of existing programs (e.g., what are the data and human resource constraints)?
3. What is the feasibility of applying such techniques on a basin-wide scale?
4. What is the availability of complementary research and how could that research be incorporated to enhance the landscape indicator work?
5. What is the potential indicator sensitivity (i.e., what are the predictive properties of these indicators)?
6. What is the applicability of specific landscape indicators for determining endpoints (i.e., ecological measurements) in ecosystems of the Great Lakes region?

This project provides several examples and a practical discussion that is designed to specifically address ecosystems in the Great Lakes Basin and its sub-basins. The maps and discussion topics in this chapter describe the differences in landscape conditions among watersheds, and the contextual background required to address the following ecosystem characteristics:

1. Areal extent of forests, wetlands, and other natural areas
2. Forest-adjacent, wetland-adjacent, and other natural areas adjacent to land cover and land use

3. Proximity of forests, wetlands, and other natural areas to anthropogenic stressors, including agriculture, urban development, and roads

4. Potential effects of anthropogenic stressors and "natural" land cover types in the vicinity of forests, wetlands, and other natural areas as they relate to

   a. Ecosystem structural characteristics

   b. Plant and animal habitat vulnerability

   c. Water quality

This project also describes landscape composition and pattern, and how such distributions may affect key ecological processes (e.g., those processes that govern the flow of energy, nutrients, water, and biota through time and space). Successful mapping of the composition and pattern of landscape conditions produce outputs that can be used to identify and characterize landscape vulnerability (i.e., risk of degradation as a result of disturbances), such as those disturbances that are directly and indirectly associated with natural and human-induced stressors (USEPA 2003). Broad-scale disturbances described in this project include those that may result in wetland ecosystem degradation as a result of fragmentation, agricultural and urban development, and hydrologic alteration in or on the periphery of wetlands.

In addition to a description and demonstration of basin-wide landscape metrics and their applicability to developing basin-wide ecological indicators, this project provides a specific example of how to implement a broad-scale ecological assessment, using a combination of the described and utilized remote sensing, GIS, and field-based techniques.

4.1.1 Great Lakes Landscape Context and Background

The entirety of the Laurentian Great Lakes can be conceptualized as, and is in fact, an ecological system (i.e., an ecosystem) that is comprised of five large lakes (i.e., Lake Superior, Lake Michigan, Lake Huron, Lake Erie, and Lake Ontario), Lake St. Clair, and their connecting channels, along with the land that drains into the lakes. The lakes are bordered to the north by the Canadian province of Ontario, and to the south by eight U.S. states (Minnesota, Wisconsin, Illinois, Indiana, Michigan, Ohio, Pennsylvania, and New York). The Great Lakes Ecosystem (hereafter, the Great Lakes or the Great Lakes Basin) forms the largest aggregated surface water body system on Earth, and comprises approximately 20% of our planet's surface water. The polar ice caps are the only other area that contains more fresh water, and the fresh water at the poles is predominantly frozen and biologically unavailable. The Great Lakes are, therefore, a major ecological contributor to the biosphere (e.g., in support of regional climate and migratory wildlife), and have also been of tremendous economic benefit to humans since

European settlement of the Great Lakes region in the seventeenth century. Covering approximately 250,000 square kilometers and draining a watershed area of approximately 500,000 square kilometers, the "freshwater seas" of the Great Lakes hold an estimated 5.7 quadrillion liters of water, which is approximately 80% of the requirements for annual water supply in the United States (USEPA 2004a). Spread evenly across the contiguous 48 states, the Great Lakes' water would be 2.9 meters deep (GLIN 2004).

Lake Superior is the largest and the deepest of the five Great Lakes and could hold the water of the four other lakes combined. Lake Michigan is located entirely within U.S. territory and is the second-deepest of the Great Lakes. Lake Huron is bounded by the lower peninsula of Michigan and Ontario, with the Georgian Bay comprising a large proportion of its water volume. The St. Clair River, Lake St. Clair, and the Detroit River connect Lake Huron to Lake Erie, which is the shallowest of the Great Lakes. Lake Erie is also the warmest of the Great Lakes, bounded by the agriculturally dominated landscapes of northern Ohio and southern Ontario. The 56-kilometer long Niagara River connects Lake Erie with Lake Ontario, sending approximately 2 million liters of water per second over Niagara Falls, through the St. Lawrence River to the Atlantic Ocean, approximately 1,600 kilometers downstream (GLIN 2004). The approximate annual outflow of water from the Great Lakes accounts for less than 1% of their total volume (Government of Canada and GLNPO 1995).

The Great Lakes are an integration of aquatic, wetland, and terrestrial ecosystems, which, despite their large size, are subject to multiscale effects of chemical and physical changes in the contributing area of the basin (Schlesinger and Bernhardt 2013; Bailey 2009). The Great Lakes Watershed contains approximately 30,000,000 million people on the American side (approximately 10% of the current U.S. population) and approximately 9,000,000 million people on the Canadian side (approximately 31% of Canada's population). Because the human population of the Great Lakes has steadily increased over time, human-induced chemical and physical disturbances have increased, particularly during the past 70 years. Humans have also recognized the beauty and commodity value of the Great Lakes but have frequently ignored the fragile composition of the entire ecosystem. However, more recent conservation and restoration efforts of aquatic, wetland, and terrestrial ecosystems (Mitsch and Jorgensen 2004) have refocused many on a new way of respecting and transforming this still-beautiful region of the world. Among the many chemical and physical disturbances in the Great Lakes, many involve hydrologic alterations that may cause increased runoff of soil, fertilizers, and pesticides from agricultural areas, or storm water runoff from residential and commercial areas. The large surface area of open water in the Great Lakes also makes them vulnerable to atmospheric deposition of pollutants by precipitation, particulates, or dust (GLIN 2004; USEPA 2004a), thus entering the flow of surface water, and especially affecting wetlands of the Great Lakes (Gorham 1987).

Ambient natural conditions that exist within the Great Lakes, such as climate, topography, physiochemical characteristics of the underlying geology, and hydrologic conditions all integrate and determine the biota of any particular location in the region. Thus, many of the coastal and inland regions contain wetlands, particularly in areas that are relatively flat, where soils are of relatively fine particle sizes, have soils with a high proportion of clay particles, and have relatively slow throughflow of water from upland areas to lake open water (Linsley and Franzini 1979). Consequently, the vegetation of these wetlands is dominated by hydrophytic plants, which are adapted to anoxic soil conditions, consequently providing specialized habitat for animals that are adapted to animal foraging, breeding, resting, or living within wetlands (Costlow et al. 1960; Vernberg and Coull 1981; Blom et al. 1990). Thus, wetlands are a big focus of this project and are among the high-priority focus areas of the Great Lakes, which can serve important ecological, economic, and societal roles in the overall functioning of the ecosystem, often referred to as wetland services or wetland functions (Costanza 1980). Wetland ecosystems, however, are a relatively small subset (by number and area) of the Great Lakes region, and most large landscapes, but owing to their relative rarity, their specialized ecological functions and human services are particularly precious and important to conserve and restore. A subset of wetland ecosystems is coastal wetlands, which consist of a narrow margin (e.g., within approximately 1 kilometer of the coastline) along limited lengths of the Great Lakes shoreline. Coastal wetlands may also be referred to as fringe wetlands, drowned river mouths, or coastal marshes and these typically extend no further than a few kilometers inland (Keough et al. 1999). Many coastal wetlands are concentrated within the large bays of the Great Lakes, such as Saginaw Bay and Green Bay, or in other smaller inlets, with many occurring at the mouths of rivers that flow to the Great Lakes. A large number of smaller areas of coastal wetlands occur in all of the Great Lakes, providing the same wetland functions and human services as the larger coastal wetlands, albeit at a finer scale and intensity.

The wetlands of the Great Lakes function as corridors of resting, breeding, and foraging habitat for birds (Prince et al. 1992). Many species of fish, amphibians, and invertebrates are full-time residents of the Great Lakes wetlands, with a subset of these species dependent on wetlands for critical portions of their life cycle. Wetlands are one of the most biologically diverse and productive ecosystems of the world (Mitsch et al. 2009). Thus, the plant communities within wetlands of the Great Lakes are a large contributor to the biological diversity and productivity of the planet. In addition to providing a desirable habitat for animals and plants, vegetational communities in wetlands help to stabilize the soil in which they grow and thus also reduce soil erosion in the basin. As a result of slowing the flow of surface water and uptake (or accumulation) of water and its constituents, wetlands in the Great Lakes, as well as elsewhere, can also provide flood control, amelioration of point and nonpoint source pollution depending upon the position of the wetland in the watershed, the types of vegetation within the wetland, and characteristics of input of water and constituents to the wetland (Government of Canada and GLNPO 1995).

Because of their relative rarity and minor portion of the overall Great Lakes Basin landscape (Figure 4.1), wetlands have been particularly impacted by the conversion of land cover within and adjacent to wetlands (Dahl 1990; Dahl and Johnson 1991). A large majority of land, aside from this minority of wetland area, in the Great Lakes Basin are either water, forest, grasslands, agriculture, or urbanized areas. Many of the direct effects (e.g., draining of wetlands and conversion of wetlands to farmland) and indirect effects (e.g., increased human population or construction of roads near wetlands) of wetland conversion over the past years are understood to have resulted in general loss and degradation of wetlands by altering their hydrology, changes in water chemistry, and reduction of the biological diversity of the plant and animal communities within these wetlands (Ball et al. 2003). Ecological disturbance theory suggests that the intensity and duration of

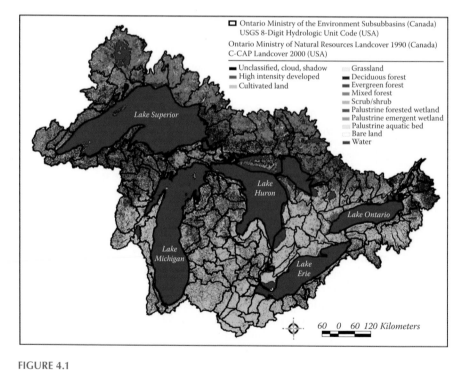

FIGURE 4.1
A synoptic view of land cover in the Great Lakes Basin. This detailed remote sensing product is a merged classification of satellite imagery from both the U.S. National Oceanographic and Atmospheric Administration's Coastal Change Analysis Program (C-CAP) land cover (https://coast.noaa.gov/digitalcoast/tools/lca, checked January 25, 2017) and Canada's Ontario Ministry of Natural Resource's land cover (http://geogratis.gc.ca/api/en/nrcan-rncan/ess-sst /fb44cd63-deb3-5efb-a07c-818e36db4c89.html, checked January 25, 2017).

such disturbances may be the key factors in the loss of ecosystem integrity, that is, the capability of an ecosystem to persist following the disturbance event (Connell and Slatyer 1977; Rapport 1990; Keddy et al. 1993; Opdam et al. 1993). As the severity, frequency, or duration of wetland disturbances increases, the survival of the plants and animals of the ecosystem are also likely to decline and one of the many observable mechanisms (Odum 1985) of the process of ecosystem degradation is the spread of nonnative (i.e., invasive) species or native opportunistic species, which heavily impact wetland ecosystems. Such losses of plant biological diversity in wetlands of the Great Lakes have been generally described (Stuckey 1989), and are demonstrated to often be an indirect effect of land cover or land use changes on the periphery of wetlands, or within these wetlands.

4.1.1.1 Landscape Characterization of Biophysical Conditions in the Great Lakes

The Great Lakes Basin (United States and Canada) was mapped using the landscape ecology approach, interpreted among the 8-digit Hydrologic Unit Codes (HUCs) for the U.S. portion of the basin and among hydrologic sub-subdivisions for the Canada portion of the basin (Figure 4.2). Because of the importance of the coastal wetlands to the Great Lakes Ecosystem and

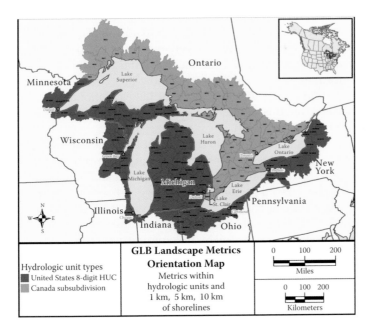

FIGURE 4.2
Basin map for the Great Lakes Ecosystem, which can be utilized to reference when viewing and interpreting geographic locations of the landscape metric maps.

decision makers, a narrow strip of area on the perimeter of the Great Lakes coastline (where coastal wetlands exist) was also mapped, separately, and a number of the landscape characteristics within this relatively small portion of the landscape was quantified and mapped. Consequently, three coastal regions were selected to analyze relevant landscape metrics in coastal zones, in addition to the entire hydrologic unit. This chapter reports metrics for solely the 1-kilometer coastal region and the full hydrologic unit to demonstrate the multi-scale approach and for ease of metric comparison between the two scales of analyses. For reference, all three coastal zones are described and depicted in the overall map of the basin, for reference, as follows:

1. A 10-kilometer coastal region, most likely encompassing all of the coastal wetlands in the basin, and a large portion of the inland landscape that may influence these coastal wetlands
2. A 5-kilometer coastal region, encompassing most all of the coastal wetlands in the basin, and a moderate portion of the inland landscape that may influence these coastal wetlands within the basin
3. A 1-kilometer coastal region, encompassing most, but not all, of the coastal wetlands in the basin, and the closely adjacent portion of the inland landscape that may influence these coastal wetlands

Although the 1-kilometer coastal region may not entirely capture all of the coastal wetlands within the basin, it is most useful for inferring the potential for disturbance of some of the landscape metrics that describe land cover that can directly affect coastal wetlands (e.g., road density and agricultural land cover metrics). The 1-kilometer coastal region is also included to provide information as per the recommendations of the State of the Lakes Ecosystem Conference and USEPA's Great Lakes National Program Office, which heretofore have assessed the condition of the Great Lakes Basin within the 1-kilometer coastal region.

Each of the coastal regions where landscape metrics are reported is also divided among the different hydrologic units of the Great Lakes Basin so that the calculations can be easily viewed and compared among them. Because the relatively narrow coastal regions are indistinguishable at the broad scale in printed formats (such as this book), and thus very difficult to portray using a standard full-basin map, each of the metrics for coastal regions (where applicable) is reported by coloring the full hydrologic unit associated with that length/ width of coastal area so that one can easily see the value within the narrower coastal region to which the HUC/value refers. Therefore, all maps of coastal and full hydrologic unit metrics are directly comparable, but you should pay close attention to the legend to identify the scale of analysis, to which the map refers (Figure 4.3). The maps in this project describe landscape metrics within the full hydrologic unit and the 1-kilometer coastal region. The metrics in the 1-kilometer coastal region are thus directly comparable, and additional

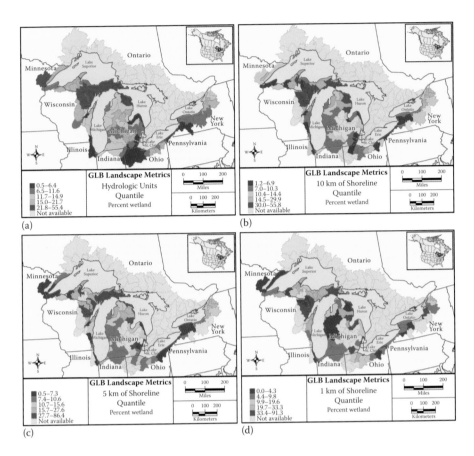

FIGURE 4.3
Regions within each of the (a) full hydrologic unit code (HUC); (b) 10-kilometer coastal zone; (c) 5-kilometer coastal zone; and (d) 1-kilometer coastal zone regions of the Great Lakes Basin (shown here solely for the U.S. side of the watershed) were mapped throughout the Basin (in these examples, for percent of watershed area that is wetland). Because the relatively narrow coastal zone regions are indistinguishable in the full basin map view here, each of the metrics here and in this chapter are reported by coloring the full hydrologic unit associated with that coastal zone width, to aid viewing and comparison among HUCs.

analyses within the 5-kilometer and 10-kilometer coastal regions are also possible (although not included in this chapter) for comparison. It is most useful to select a few key metric maps for comparison and explore the differences in selected metrics among watersheds, which are most relevant to your interests.

4.1.1.2 Measuring Landscape Ecological Vulnerability in the Great Lakes

As we know, ecosystems of the Great Lakes Basin are vulnerable to loss or degradation as a result of the interaction of naturally occurring conditions and human activities (Table 4.1). Ecosystems that are degraded as a result

TABLE 4.1

Environmental Conditions in the Great Lakes Basin and Some of the Potential
Effects upon Ecosystems with a Special Focus on Wetlands

Great Lakes Basin Environmental Condition	Potential Coastal Wetland Effect(s)
Adjacent urbanization	Peak flows of runoff from paved urban areas may rapidly pulse through wetland and increase the amount of metals, oils, salts, or other contaminants into, or flowing out of, wetlands to open lake areas
Change in magnitude or duration or frequency of water levels	Changed competitive or successional processes that may result in changed species diversity in fish, amphibian, bird, plant, or other community structures
Change in wetland vegetation, e.g., change in proportion of wetland open water and emergent vegetation	Loss of optimum habitat for some species of fish, waterfowl, and other marsh birds
Chemical/oil spill	Death of wetland organisms
Dredging	Deepening water and removal of sediments can result in loss of wetland habitat
Early ice breakup, early peaks in spring runoff, change in the timing of stream flow, and increased intensity of rainstorms	Fewer viable breeding sites, especially for amphibians, migratory shorebirds, and waterfowl; northern migratory species (e.g., Canada geese) winter further north; increased flooding frequency in coastal areas
Habitat loss and fragmentation	Decrease in the available aquatic habitat for organisms, especially affecting species with limited dispersal capabilities (e.g., amphibians and mollusks)
Mechanical clearing of wetland vegetation	Creation of impassable areas for some species, thus isolating populations and increasing likelihood of extirpation
Over-harvesting of resources	Depletion of recreationally or commercially valuable species
Reduced summer water levels	Reduction in the total area of wetlands, resulting in poorer water quality and less habitat for wildlife
Removal of tree cover and shoreline vegetation	Increased runoff into wetland from adjacent land
Runoff and pollutants from agricultural areas, sewage treatment outflows, stormwater outputs, urbanized areas, industrial outfalls, and other sources in watershed	Increased loading of nutrients, sediments, and toxic chemicals in downstream wetlands; reduced water clarity
Shoreline modification; wetland filling or drainage	Physical destruction or reduction in protection of coastal regions to erosion
Species invasion and spread (e.g., carp, zebra mussel, common reed, purple loosestrife)	Feeding, spawning, and nesting behavior of animals may interfere with plant photosynthesis/growth; non-native animals may prey upon native animal species or outcompete them for food and habitat; plants may not provide suitable forage, nesting, reproduction
Storms and seiches	Damage to vegetation due to high winds and waves

of conditions within the Great Lakes Basin may continue to function, but at a reduced functional level. Not all ecosystems remain after these functional changes occur, with some losing their ecological functions quickly, and others ceasing to function altogether (i.e., ecosystem loss, often from land development). Ecosystems may flourish in conditions that fluctuate in their conditional state; for example, some ecosystems depend on periodic changes in hydrology, which tends to influence the diversity of plants, which in turn supports ecosystem-independent animal habitat (e.g., in wetlands or wet forest types). Thus, periodic disturbances may allow for the formation of relatively small, interconnected metapopulations, where gene flow between plant patches or wetlands maintains the genetic diversity that might otherwise decline in relatively large inbred populations. When such populations become unable to bridge the gaps between populations, at the advanced stages of patch isolation, entire populations may become locally extinct (Opdam 1990). Water-level fluctuations also promote the interaction of aquatic and terrestrial ecosystems, and can result in higher quality habitat and increased productivity within and among ecosystems.

In wetlands, environmental changes that can directly influence ecological condition (e.g., as a result of dredging, filling, draining, and species invasion) and are therefore easier to pinpoint than indirect environmental changes. Direct environmental changes in wetlands are often human-induced, highly visible, and can result in rapid changes to wetlands. Indirect environmental changes are often less pronounced, potentially causing changes in wetland function and vegetation communities over a longer period of time. Indirect environmental changes are physically removed from a wetland, and thus it may be difficult to pinpoint the exact source of the environmental change. Indirect environmental changes include urban and agricultural runoff. Indirect environmental changes are relatively difficult to control, due to their diffuse and variable sources (Environment Canada 2002). Human-induced environmental change factors analyzed in this project are based on previously observed positive correlations between ecosystem degradation and amount of land cover conversion during road construction, road maintenance, and other human activities (e.g., Connell and Slatyer 1977; van der Valk 1981; Ehrenfeld 1983; Johnston 1989; Scott et al. 1993; Johnston 1994; Jenning 1995; Poiani and Dixon 1995; Wilcox 1995; Ogutu 1996; Stiling 1996; Lopez et al. 2002).

One of the challenges of using remote sensing for landscape ecology is that the technology of remote sensing allows for massive amounts of information to be gathered across vast areas, and thus contributing to a vast array of new ways to view and interpret the configuration of land cover and land use we see across the diverse region is possible. Because of the vastness of certain landscapes, such as the Great Lakes, a method for synoptically assessing and summarizing landscape and watershed condition information is necessary, and an adaptation of the 1990s (Jones et al 1997) landscape atlas format, which presents a number of key maps that used colored legends to indicate ranking of particular landscape metrics and indicators, was selected for analyses of the vast Great Lakes area (Figure 4.1). Accordingly, the methods utilized here in this chapter have been used to provide digital interpretations (sometimes referred to as landscape "metric browsers"), increasing the ability of this tremendously large remote sensing data set for those in the Great Lakes, in a manner that is "user friendly." There are several examples of this method provided in this book, namely the projects in the Great Lakes, Missouri River and Mississippi River Basins, and the Ozark Mountains, which all utilize to some degree the landscape atlas format to assimilate and portray a tremendous amount of land cover, landscape metric, and landscape indicator data.

Base land cover data for the analyses in the Great Lakes project are the merged data sets from the U.S. National Oceanographic and Atmospheric Administration's 2000s Coastal Change Analysis Program (C-CAP) land cover (https://coast.noaa.gov/digitalcoast/tools/lca, checked January 25, 2017) for the U.S. side of the Great Lakes Basin, and Canada's Ontario Ministry of Natural Resource's 1990s land cover data sets (http://geogratis.gc.ca/api/en /nrcan-rncan/ess-sst/fb44cd63-deb3-5efb-a07c-818e36db4c89.html, checked January 25, 2017) for the Canada side of the Great Lakes Basin (Figure 4.1). The combined remote sensing analysis for the entire Great Lakes Watershed is presented in this chapter in a viewable format that allows the user to notice subtle differences among watersheds, at different scales of analyses, with mapping and interpretation of landscape scale ecological metrics that are possible among hydrologic units, and to effectively demonstrate the differences between scales of analyses for coastal regions, we discuss solely the 1-km and full HUC scaled metrics in this chapter. The maps in this case study present key ecological metrics that are necessary to determine the current condition of the Great Lakes Basin, and they provide a baseline for analyses of future landscapes in the Great Lakes Ecosystem.

The vastness of the area involved in this synoptic ecological assessment of the entire Great Lakes Basin should make one consider the use of remote sensing and GIS techniques to measure the potential for ecological disturbance in other large regions, and trigger the thought of what the equivalent work would be if there were less than complete coverage and techniques for determining land cover classes as has been accomplished for these data sets. The binational efforts to assess landscape scale disturbances in the Great Lakes Basin (both the United States and Canada) utilized available multispectral satellite data processing, spatial data set merging, and GIS modeling necessary for their respective land cover data sets, which were necessary for the determination of the landscape ecology metrics outlined below.

Metrics in the Great Lakes that are most relevant to the overall landscape ecological condition, and were discernable at the synoptic level using Landsat-based remote sensing, are:

1. General Landscape Characteristics: The percent of selected land cover types observed throughout the Great Lakes Basin is provided

to acquaint the user of this browser with the extent of land that has been assessed by this project and the distribution of these land cover types throughout the GLB.

2. Sediment Flowing to Coastal Wetlands (SOLEC ID: 4516 [SOLEC 2000, 2009]): The amount of sediment and other runoff constituents within surface waters that flow from headwaters to coastal wetlands in the coastal watersheds of the Great Lakes Basin.

3. Sediment Available for Coastal Nourishment (SOLEC ID: 8142 [SOLEC 2000, 2009]): The amount of sediment in the coastal zone of the waters of the Great Lakes Basin.

4. Urban Density (SOLEC ID: 7000 [SOLEC 2000, 2009]): The area of urban land in the Great Lakes Basin.

5. Land Conversion (SOLEC ID: 7002 [SOLEC 2000, 2009]): The amount of change from natural land cover types to human land use types in the Great Lakes Basin.

6. Habitat Adjacent to Coastal Wetlands (SOLEC ID: 7055 [SOLEC 2000, 2009]): The amount and characteristics of habitat adjacent to coastal wetlands in the Great Lakes Basin.

7. Habitat Fragmentation (SOLEC ID: 8114 [SOLEC 2000, 2009]): The amount of plant and animal habitat dissection in the Great Lakes Basin.

The spatial configuration of ecosystems (i.e., size, shape, and interspersion within the larger landscape) is an important consideration, since larger ecosystems may be relatively more likely to persist in the face of environmental changes. Wetlands of various sizes, for example, attract different species, and a range of sizes may increase the diversity of habitat types across a broad area. For example, some birds (e.g., black tern, Forster's tern, and short-eared owl) may require a sufficiently large size before they will make use of it for nesting (Environment Canada 1998). Mitsch and Gosselink (2009) have described wetlands as spatially and temporally dynamic habitats, and thus the boundaries of wetlands could be affected by the combined geological and hydrological processes associated with erosion and deposition, changing biological processes in the process. Wetland size and proximity metrics used in this chapter are based on previously observed trends regarding the effects of patch size, patch shape, and the interspersion of ecosystems within the broader landscape for specific taxa, in many different regions (e.g., MacArthur and Wilson 1967; Simberloff and Wilson 1970; Diamond 1974; Forman et al. 1976; Pickett and Thompson 1978; Soule et al. 1979; Hermy and Stieperaere 1981; van der Valk 1981; Simberloff and Abele 1982; Harris 1984; McDonnell 1984; Moller and Rordam 1985; Brown and Dinsmore 1986; Dzwonko and Loster 1988; Gutzwiller and Anderson 1992; Opdam et al. 1993; Hamazaki 1996; Kellman 1996; Bastin and Thomas 1999; McIntyre and

Wiens 1999a, 1999b; Twedt and Loesch 1999; Jones et al. 2000; Lopez et al. 2002; Lopez and Fennessy 2002).

The spatial configuration of ecosystems within the larger landscape is also important because ecosystem vulnerability (i.e., the risk of loss or degradation) can be initially evaluated by investigating these spatial interrelationships. Measurement of the spatial configuration of ecosystem size, shape, spacing, proximity to non-natural land cover types, and variations of these metrics are important because these metrics may foretell the likelihood that a particular ecosystem will rebound after a disturbance. That is, an a priori understanding of ecosystem characteristics, and specific landscape metrics can be used to address ecological endpoints that predict habitat degradation as a result of ecosystem destruction, fragmentation, or degradation.

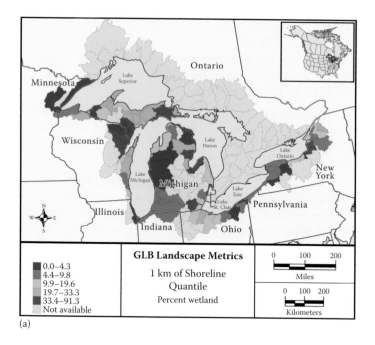

(a)

FIGURE 4.4
Percent wetland of 1-kilometer coastal region among all watersheds in the Great Lakes Basin within (a) U.S. percent coastal wetland is calculated by dividing the number of wetland land cover cells in the coastal region of each watershed (i.e., the reporting unit) by the total number of land cover cells in the reporting unit minus those cells classified as water. This measurement has potential for measuring and comparing wetland contribution among watersheds and may be used to indicate potential for wetland removal or reduction in the amount of pollutants entering the Great Lakes. The relative extent of wetlands may also be developed into a quantitative indicator of habitat for a wide variety of plant and animal species. (From the United States Hydrologic Units [8-digit HUCs], United States Coastal Change Analysis Program [CCAP], Canada Hydrologic Units [Subsubdivisions], and Ontario Ministry of Natural Resources Data Set.)   *(Continued)*

4.1.2 Determining Terrestrial Conditions

4.1.2.1 *Areal Extent of Great Lakes Wetlands*

The extent of wetlands in coastal areas (i.e., wetlands within 1 kilometer of the coastline of the Great Lakes) is shown in Figure 4.4. The differences between wetland areal coverage among watersheds and coastal region areas can be used to interpret other metrics and to prioritize all of the Great Lakes wetlands in a more detailed manner, by contrasting the analyses at various scales.

Prior to European settlement, the extent of wetlands in the Great Lakes Basin spanned large areas from the western edge of Lake Erie, across Ohio and Indiana, and covering the southern portion of the Province of Ontario.

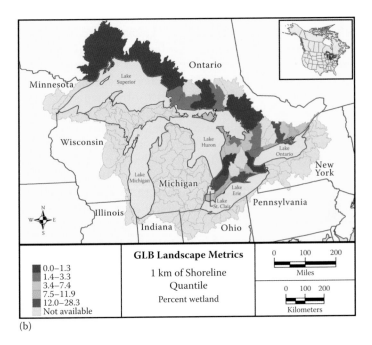

(b)

FIGURE 4.4 (CONTINUED)
Percent wetland of 1-kilometer coastal region among all watersheds in the Great Lakes Basin within (a) the United States and (b) Canada. Percent coastal wetland is calculated by dividing the number of wetland land cover cells in the coastal region of each watershed (i.e., the reporting unit) by the total number of land cover cells in the reporting unit minus those cells classified as water. This measurement has potential for measuring and comparing wetland contribution among watersheds and may be used to indicate potential for wetland removal or reduction in the amount of pollutants entering the Great Lakes. The relative extent of wetlands may also be developed into a quantitative indicator of habitat for a wide variety of plant and animal species. (From the United States Hydrologic Units [8-digit HUCs], United States Coastal Change Analysis Program [CCAP], Canada Hydrologic Units [Subdivisions], and Ontario Ministry of Natural Resources Data Set.)

It is estimated that two-thirds of Great Lakes wetlands have been lost since European settlement. Many of these areas have been drained or reclaimed for land development, farmland, and harbors along the coast of the Great Lakes, which may contribute to the degradation of wetlands. Other impacts to wetlands include suburbanization, dam construction, stream alteration, and the construction of flood control structures, which also alter the hydrology of contributing watersheds (Cox and Cintrón 1997).

Between the 1780s and the 1980s, the largest reductions of wetlands occurred in Ohio (USEPA 2001b). Urban development along the shores of the Great Lakes generally reflects the history of human decision-making processes that necessitated safe and efficient harbors for the distribution of natural resources, such as timber and mineral ores. As a result of these decisions, the areal extent of wetlands has been dramatically reduced by the conversion to, and to some extent by, the indirect effects of urban and agricultural land use (USEPA 2002).

The loss of forested land in the Great Lakes is apparent by an analysis of the proportion of remaining forested lands (Figure 4.5); this is forestland that has been predominantly replaced by cropland and pasture (Figure 4.6). The loss of forestland has also been slowed and to some extent reversed in some areas, along with other wooded "shrublands" (Figure 4.7), through the processes of ecological restoration or natural regrowth. The patterns of deforestation, reforestation, and regrowth have resulted in the current patchwork of forested areas across the Great Lakes Basin, which have tremendous local-to-global influences the biophysical environment and society (Bailey 2009), such as with the regulation of carbon fluxes (Tang et al. 2009). The pattern of forested areas in the Great Lakes has a profound impact on the habitat characteristics of those areas for vertebrates and invertebrates, which can be explored using several metrics of connectivity (Figure 4.8), forest edge (Figure 4.9), and fragmentation (Figure 4.10).

4.1.2.2 Proximity of Land Cover and Land Use to Wetlands

As with most coastal areas around the world, the coastal region of the Great Lakes has been an attractive location for development during the history of settlement and expansion of societies into the region. The shorelines are a focus of human activities because they are near water, which provides unique transportation functions, resources for manufacturing, recreational opportunities, residential uses, and drinking water resources. The transportation services in combination with the close proximity to productive farmland, raw materials, and an ever-growing inland infrastructure makes the coastal areas an unparalleled area to economically exploit. Thus, there may be conflicts between preserving the remaining wetlands and developing these areas for additional commercial and societal needs. Wetland areas that are close to urbanization (Figure 4.11) or

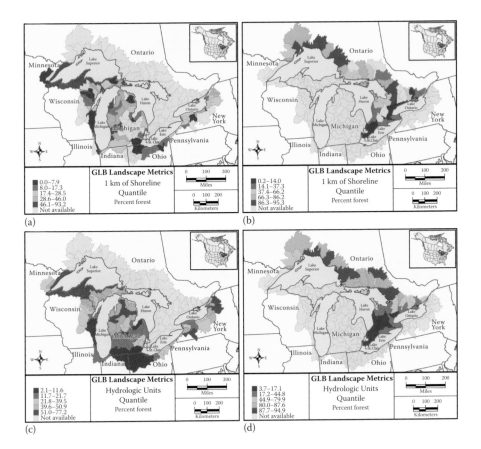

FIGURE 4.5
Percent forest within/among the 1-kilometer coastal regions of (a) the U.S. and (b) the Canada watersheds of the Great Lakes Basin, and within/among the full watersheds of the (c) U.S. and (d) Canada portions of the Great Lakes Basin. The percentage of forestland cover is calculated by dividing the number of forestland cover cells in the reporting unit by the total number of land cover cells in the reporting unit minus those cells classified as water (i.e., total land area). Forests may remove or reduce the amount of pollutants entering streams and lakes. Forests also provide habitat for a wide variety of plant and animal species. (From the United States Hydrologic Units [8-digit HUCs], United States Coastal Change Analysis Program [CCAP], Canada Hydrologic Units [Subsubdivisions], and Ontario Ministry of Natural Resources Data Set.)

dense human population centers (Figure 4.12) may be sensitive natural areas and affected by human land use associated with urban and suburban activities.

Another example of the effects of ecosystem disturbance is an increased expansion of invasive or opportunistic plants into landscape gaps (e.g., within and between wetlands), which may be the result of increased land cover fragmentation (Forman 1995). The patch dynamics (i.e., either increases

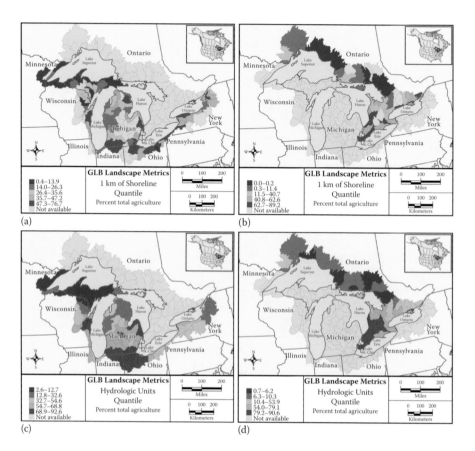

FIGURE 4.6
Percent total agriculture (crops and pasture) within/among the 1-kilometer coastal regions of
(a) the United States and (b) Canada watersheds of the Great Lakes Basin, and within/among
the full watersheds of (c) the U.S. and (d) the Canada portions of the Great Lakes Basin. The
percentage of all agricultural land cover is calculated by dividing the number of all agricultural land cover (i.e., pasture and crop) cells in the reporting unit by the total number of land
cover cells in the reporting unit minus those cells classified as water (i.e., total land area).
Agricultural practices typically employ fertilizers, pesticides, and other chemicals that may
be transported to streams in water runoff. The closer agriculture is to a stream the more likely
related pollutants will enter the stream. Concentrations of pollutants transported into streams
are also more likely to be higher when agriculture is closer to streams. Animals grazing on
pastures may decrease vegetation cover, possibly leading to increased runoff and erosional
soil loss, which may result in increased stream sedimentation. Livestock may also degrade
within-stream and stream-bank ecological functions by defecating in the streams and tram-
pling riparian vegetation, respectively. (From the United States Hydrologic Units [8-digit
HUCs], United States Coastal Change Analysis Program [CCAP], Canada Hydrologic Units
[Subsubdivisions], and Ontario Ministry of Natural Resources Data Set.)

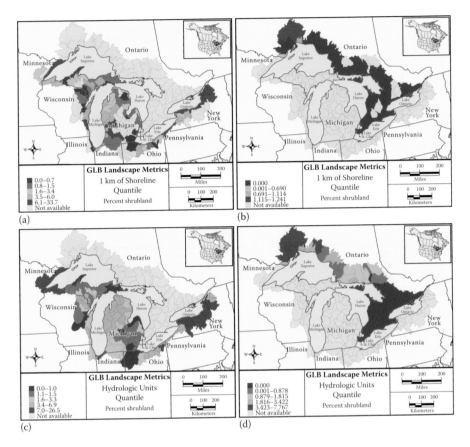

FIGURE 4.7

Percent shrubland within/among the 1-kilometer coastal regions of the (a) U.S. and (b) Canada watersheds of the Great Lakes Basin, and within/among the full watersheds of the (c) U.S. and (d) Canada portions of the Great Lakes Basin. The percentage of shrubland land cover is calculated by dividing the number of shrubland land cover cells in the reporting unit by the total number of land cover cells in the reporting unit minus those cells classified as water (i.e., total land area). (From the United States Hydrologic Units [8-digit HUCs], United States Coastal Change Analysis Program [CCAP], Canada Hydrologic Units [Subsubdivisions], and Ontario Ministry of Natural Resources Data Set.)

or decreases in extent) of invasive and opportunistic plant species (Nash et al. 2005) in disturbed areas of the Great Lakes may be facilitated by the extent and intensity of wetland patch disturbance that results from human fragmentation of the landscape, resulting in hydrologic alteration (e.g., from road construction, where roads cross streams and rivers). Because species-level assessments may not be possible using satellite, or other coarse-scale remote sensing data (i.e., spatial or spectral resolution data), it may be necessary to map invasive or opportunistic species using finer-scale remote sensing data.

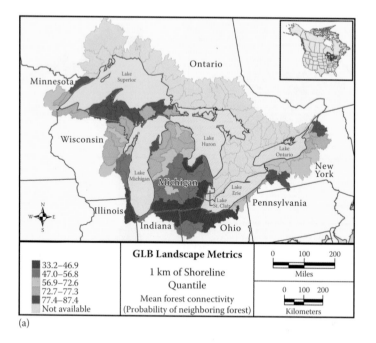

(a)

FIGURE 4.8

Forest connectivity (probability of neighboring forest) within/among the full watersheds of the (a) U.S. Probability of a forest cell having a neighboring forest cell is calculated using a moving 270-meter-square window (9 pixels × 9 pixels) across the land cover. The boundaries between all pixel pairs where at least one pixel is forest are examined in the window. The metric is the number of boundaries where both pixels are forest, divided by the total number of forest boundaries (regardless of neighbor land cover type). This metric gives a measure of how well the forest is connected within the window sample area, with high values being better connected than low values. (From the United States Hydrologic Units [8-digit HUCs], United States Coastal Change Analysis Program [CCAP], Canada Hydrologic Units [Subsubdivisions], and Ontario Ministry of Natural Resources Data Set.)   (*Continued*)

Data that describes land cover and land use within the vicinity of, or directly adjacent to, wetlands may be important as indicator information about the level of disturbance upon wetlands (as with other ecosystem types, perhaps). For example, paved surfaces (e.g., roads; Figure 4.13) increase the impermeability (Figure 4.14) of land surfaces and may increase the amount of runoff (May et al. 1997) to streams, lakes, and wetlands, and potentially increase the transport of road salts or other chemicals from paved surfaces (e.g., trace metals and hydrocarbons). Roads also fragment habitat and may act as barriers to animal movement (e.g., amphibians or large mammals).

Land use in a particular watershed may also have a significant influence on the flow of runoff and sediments toward coastal areas, and may be indicative of the amount of runoff that is intercepted by wetlands in that

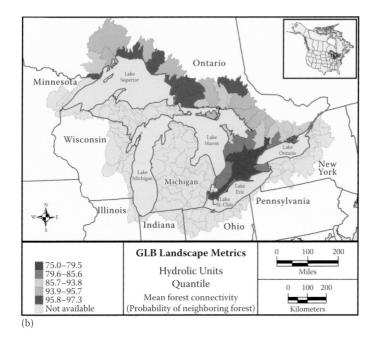

(b)

FIGURE 4.8 (CONTINUED)
Forest connectivity (probability of neighboring forest) within/among the full watersheds of the (a) U.S. and (b) Canada portions of the Great Lakes Basin. Probability of a forest cell having a neighboring forest cell is calculated using a moving 270-meter-square window (9 pixels × 9 pixels) across the land cover. The boundaries between all pixel pairs where at least one pixel is forest are examined in the window. The metric is the number of boundaries where both pixels are forest, divided by the total number of forest boundaries (regardless of neighbor land cover type). This metric gives a measure of how well the forest is connected within the window sample area, with high values being better connected than low values. (From the United States Hydrologic Units [8-digit HUCs], United States Coastal Change Analysis Program [CCAP], Canada Hydrologic Units [Subsubdivisions], and Ontario Ministry of Natural Resources Data Set.)

watershed. The capability of such wetlands to accumulate, transform, or store pollutants that are transported in the runoff from the inland areas of the watershed is an important mechanism for maintaining and improving the water quality of the Great Lakes. Wetlands that are adjacent to other habitats and that provide connections between other habitats in the watershed are also more likely to maintain their normal hydrologic regime, which may moderate the amount of water, sediment, and chemical constituents that are directly input into the open water areas of the lakes and other downstream areas. Thus, areas that are relatively more developed and intensively used for agriculture may have increased rates of runoff and sediment loading to the Great Lakes. However, if wetlands or other forested/shrubby/grassy areas are situated between upland urban or agricultural areas, and open

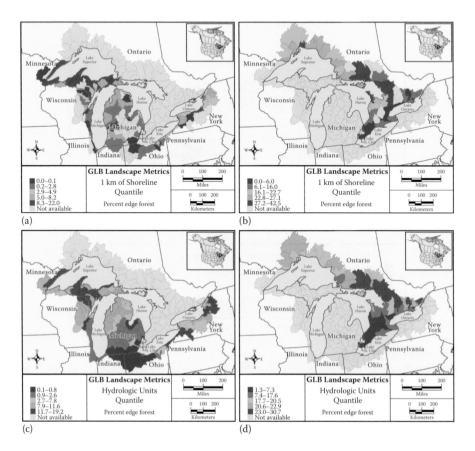

FIGURE 4.9
Percent forest edge within/among the 1-kilometer coastal regions of the (a) U.S. and (b) Canada watersheds of the Great Lakes Basin, and within/among the full watersheds of the (c) U.S. and (d) Canada portions of the Great Lakes Basin. Percentage of edge forest is calculated using a moving 270-meter-square window (9 pixels × 9 pixels) across the land cover. When the percent forest in the window is greater than 60%, but less than the window's mean forest connectivity value, the forest cell in the center of the window is classified as edge. The number of edge forest cells in the reporting unit is then divided by the reporting unit's total land area (the total number cells in the reporting unit boundary minus those cells classified as water) to derive the percentage of edge forest. Edge forest indicates largely continuous, clumped forest, which is more likely to provide increased opportunities for upland forest animal movement and upland forest plant dispersal, but may or may not provide suitable interior habitat, depending on the species. It is important to recognize that increased amounts of edge may increase the likelihood of invasive (i.e., non-native) or opportunistic (i.e., native) animal or plant species occurrences within a forest, and may result in decreased animal or plant diversity. (From the United States Hydrologic Units [8-digit HUCs], United States Coastal Change Analysis Program [CCAP], Canada Hydrologic Units [Subsubdivisions], and Ontario Ministry of Natural Resources Data Set.)

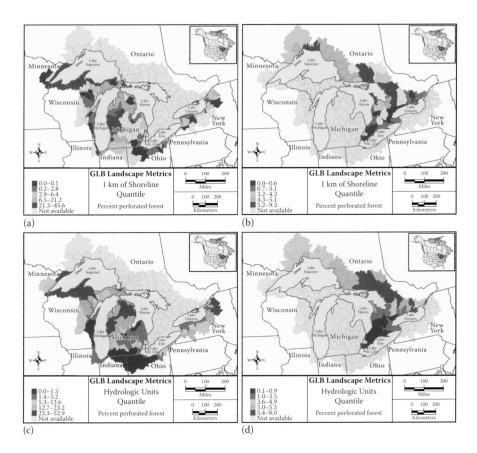

FIGURE 4.10
Percent perforated forest within/among the 1-kilometer coastal regions of the (a) U.S. and (b) Canada watersheds of the Great Lakes Basin, and within/among the full watersheds of the (c) U.S. and (d) Canada portions of the Great Lakes Basin. Percentage of perforated forest is calculated using a moving 270-meter-square window (9 pixels × 9 pixels) across the land cover. When the percent forest in the window is greater than 60%, and greater than the window's mean forest connectivity value, the forest cell in the center of the window is categorized as perforated. The number of perforated forest cells in the reporting unit is then divided by the reporting unit's total land area (i.e., the total number of cells in the reporting unit boundary minus those cells classified as water) to derive the percentage of perforated forest. Perforated forest generally consists of a patch of forest with a center non-forested area, such as would occur if a small clearing were made within a patch of forest. Perforated forest may be fragmented in this fashion to such an extent that they do not provide suitable interior habitat for some upland forest species. (From the United States Hydrologic Units [8-digit HUCs], United States Coastal Change Analysis Program [CCAP], Canada Hydrologic Units [Subsubdivisions], and Ontario Ministry of Natural Resources Data Set.)

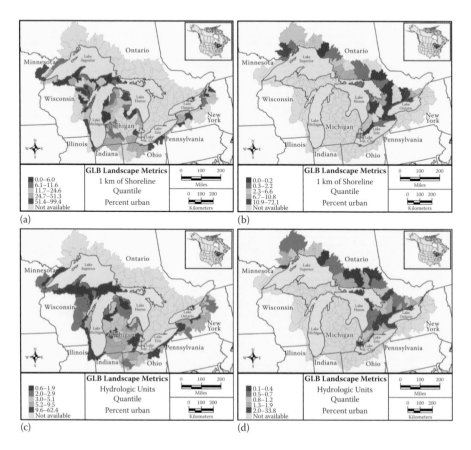

FIGURE 4.11
The percentage of urban land cover within/among the 1-kilometer coastal regions of the
(a) U.S. and (b) Canada watersheds of the Great Lakes Basin, and within/among the full watersheds of the (c) U.S. and (d) Canada portions of the Great Lakes Basin. This metric is calculated
by dividing the number of urban land cover cells in the reporting unit by the total number of
land cover cells in the reporting unit minus those cells classified as water (i.e., total land area).
High amounts of urban land indicate substantial modification of natural vegetation cover and
may affect the condition of wildlife habitat, soil erosion, and water quality in the Great Lakes
Watershed. (From the United States Hydrologic Units [8-digit HUCs], United States Coastal
Change Analysis Program [CCAP], Canada Hydrologic Units [Subsubdivisions], and Ontario
Ministry of Natural Resources Data Set.)

water, the runoff and sediment loading may be reduced to those open water
areas. It is important to also note that wetlands in close proximity to urban
or agricultural land (Figure 4.15) may be at greater risk of loss or degradation as a result of hyper-eutrophication or pollution, and their capacity to
ameliorate environmental impacts from sediment and pollutants is not lim-
itless. Wetlands that are adjacent to urban land cover (Figure 4.16) may also

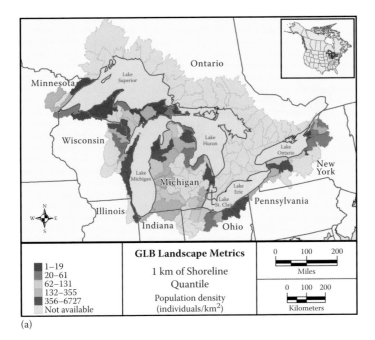

(a)

FIGURE 4.12

Human population density (individuals/km²) approximated within (a) the 1-kilometer coastal region. Population density is calculated by summing the number of people living in the reporting unit and dividing by the reporting unit area. Where census units are not completely contained within the reporting unit, population is apportioned by area. High population densities are generally well correlated with high amounts of human land uses, especially urban and residential development. Large areas of development often involve substantial modification of natural vegetation cover that may have substantial effects on wildlife habitat, soil erosion, and water quality. (From the United States Hydrologic Units [8-digit HUCs], United States Coastal Change Analysis Program [CCAP], and United States Census 2000 [c2k].)   (*Continued*)

provide poor animal habitat relative to wetlands adjacent to natural land cover, such as forests.

4.1.3 Determining Patch Characteristics

4.1.3.1 Inter-Wetland Spacing and Landscape Integration

Interconnected wetland patches function as a network (e.g., within a watershed or migratory bird flyway), and have the cumulative functional capability of all the individual wetlands. A collection of wetlands in the landscape may be particularly important for providing a vital ecological unit for some animals, while other animals may require a mixture of wetland and upland areas for different portions of their life cycle or their daily activities (e.g., a species that reproduces in wetlands and forages in upland areas). The absence of

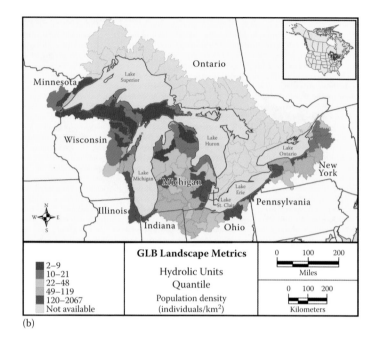

(b)

FIGURE 4.12 (CONTINUED)
Human population density (individuals/km²) approximated within (b) the full HUC of the U.S. portion of the Great Lakes Basin. Population density is calculated by summing the number of people living in the reporting unit and dividing by the reporting unit area. Where census units are not completely contained within the reporting unit, population is apportioned by area. High population densities are generally well correlated with high amounts of human land uses, especially urban and residential development. Large areas of development often involve substantial modification of natural vegetation cover that may have substantial effects on wildlife habitat, soil erosion, and water quality. (From the United States Hydrologic Units [8-digit HUCs], United States Coastal Change Analysis Program [CCAP], and United States Census 2000 [c2k].)

such wetland complexes or integrated upland and wetland conditions may completely interrupt or degrade the reproduction rates, survival rates, and overall fitness of some plant and animal species.

Fragmentation of the landscape may result in the isolation of wetlands, with the remnants of the formerly larger interconnected wetland complexes being replaced by less heterogeneous landscapes that are dominated by either agricultural land, urban or rural human habitations, and industrial land. Such conversions of wetland to other land cover types may reduce the functional capability of wetlands and may have also increased the likelihood that the remaining wetlands are further affected by the new land cover type (Tiner et al. 2002). Thus, as the general concept of ecosystem integrity describes, the capability of wetlands to continue to function and provide

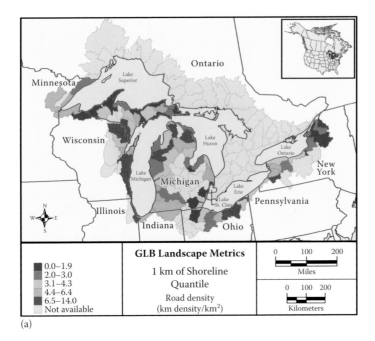

(a)

FIGURE 4.13

Road density (km road/km²) approximated within (a) the 1-kilometer coastal region. The density of roads is calculated by summing the length of roads and dividing by the area of the reporting unit. Values are reported as km of all road types (freeways, highways, surface streets, rural routes, etc.) per km². High total road densities are generally well correlated with high human population and urban development. Roads increase the impermeability of land surfaces, may increase the amount of runoff to streams and lakes, and potentially increase the transport of road salts or other chemicals from paved surfaces (e.g., trace metals and hydrocarbons). Roads also fragment habitat and may act as barriers to animals (e.g., amphibians or large mammals). (From the United States Hydrologic Units [8-digit HUCs], United States Coastal Change Analysis Program [CCAP], United States Census 2000 [c2k], and United States Roads [Wessex/GDT].) *(Continued)*

ecological services to the residents of the Great Lakes (e.g., improving and maintaining clean water; providing critical habitat for plants and animals; and shoreline stabilization and protection) is dependent upon the effects of the surrounding landscape.

Wetland connectivity is one way of measuring the fragmentation of wetlands in the Great Lakes region (Figure 4.17). A standard and uniform method for measuring wetland connectivity is to determine the probability of a wetland area cell having a neighboring wetland, using a "moving window" over a GIS data set (i.e., a 9-pixel × 9-pixel area in Figure 4.17). In this project, the boundaries between all pixel pairs, where at least one pixel is wetland, were examined using a moving window method. The resulting connectivity

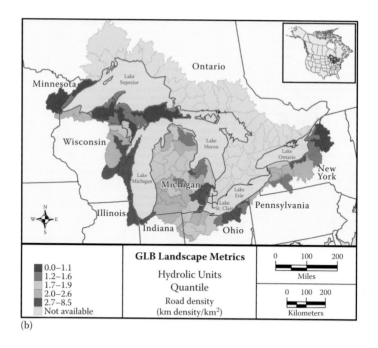

(b)

FIGURE 4.13 (CONTINUED)
Road density (km road/km²) approximated within (b) the full HUC of the U.S. portion of the Great Lakes Basin. The density of roads is calculated by summing the length of roads and dividing by the area of the reporting unit. Values are reported as km of all road types (freeways, highways, surface streets, rural routes, etc.) per km². High total road densities are generally well correlated with high human population and urban development. Roads increase the impermeability of land surfaces, may increase the amount of runoff to streams and lakes, and potentially increase the transport of road salts or other chemicals from paved surfaces (e.g., trace metals and hydrocarbons). Roads also fragment habitat and may act as barriers to animals (e.g., amphibians or large mammals). (From the United States Hydrologic Units [8-digit HUCs], United States Coastal Change Analysis Program [CCAP], United States Census 2000 [c2k], and United States Roads [Wessex/GDT].)

metric is the number of boundaries where both pixels are wetland, divided by the total number of wetland boundaries (regardless of neighbor land cover type). This metric gives a measure of how well the wetland is connected within the window sample area, with high values being better connected than low values.

The relative percentage of "perforated" wetland is another measurement of ecosystem fragmentation (Turner et al. 2001), and is calculated here by using a moving 270-meter-square window (i.e., 9 pixel × 9 pixel) across the GIS land cover data set (Figure 4.18).

When the percent wetland in the window is greater than 60%, and greater than the window's mean wetland connectivity value (Figure 4.17), the wetland cell in the center of the window is categorized as perforated (Figure 4.18).

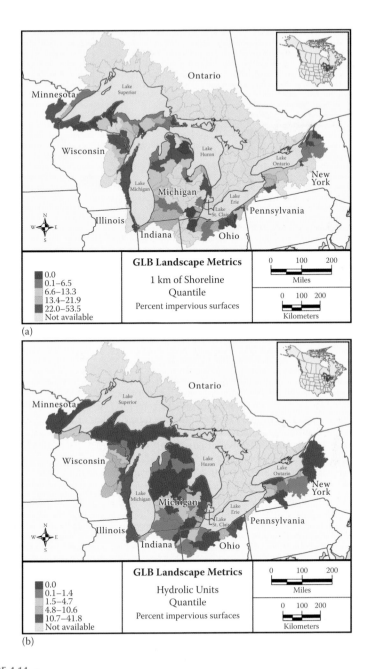

FIGURE 4.14
Percent impervious surfaces within (a) the 1-kilometer coastal region and within (b) the full HUC of the U.S. portion of the Great Lakes Basin. The percent total impervious area is calculated using road density as the independent variable in a linear regression model (see May et al. 1997) utilizing the following land cover data sources: the United States Hydrologic Units (8-digit HUCs) and the United States Coastal Change Analysis Program (CCAP).

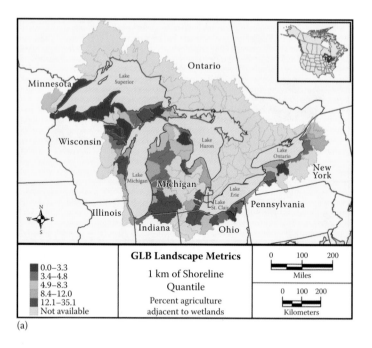

(a)

FIGURE 4.15

Percent agriculture adjacent to wetlands within (a) the 1-kilometer coastal region. The effect of runoff from areas adjacent on wetlands may be strongly influenced by agricultural land cover type immediately adjacent to a wetland. Thus, the percentage of agricultural land cover within 30 meters of wetlands is calculated by summing the total number of urban land cover cells within a single cell buffer (30-meter) zone on the perimeter of wetland patches in the reporting unit and dividing by the area of the buffer plus the area of the buffered wetland. Cells outside buffer zones for wetlands within each reporting unit are ignored. Other buffer distances may be more appropriate, depending on the runoff chemical constituent(s), flow dynamics, soil conditions, position of wetland in the landscape, and other landscape characteristics. (From the United States Hydrologic Units [8-digit HUCs] and the United States Coastal Change Analysis Program [CCAP].)   *(Continued)*

The number of perforated wetland cells in the reporting unit (i.e., a HUC or a 1-kilometer coastal zone of a HUC) is then divided by the reporting unit's total land area (i.e., the total number of cells in the reporting unit boundary minus those cells classified as water) to derive the percentage of perforated wetland. Perforated wetland generally consists of a patch of wetland with center upland area(s), such as would occur if small clearing(s) were made within a patch of wetland, or if an area of wetland contained an interior upland region. Perforated wetlands may be fragmented in this fashion to such an extent that they do not provide suitable interior habitat for some wetland species. However, the interspersion of upland and wetland conditions in perforated wetlands may provide suitable habitat for some specialized

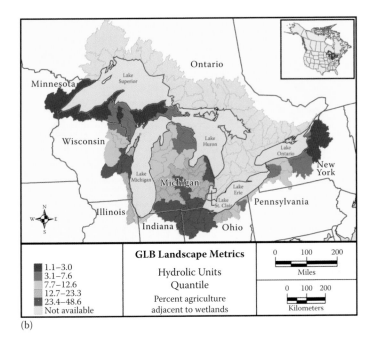

(b)

FIGURE 4.15 (CONTINUED)

Percent agriculture adjacent to wetlands within (b) the full HUC of the U.S. portion of the Great Lakes Basin. The effect of runoff from areas adjacent on wetlands may be strongly influenced by agricultural land cover type immediately adjacent to a wetland. Thus, the percentage of agricultural land cover within 30 meters of wetlands is calculated by summing the total number of urban land cover cells within a single cell buffer (30-meter) zone on the perimeter of wetland patches in the reporting unit and dividing by the area of the buffer plus the area of the buffered wetland. Cells outside buffer zones for wetlands within each reporting unit are ignored. Other buffer distances may be more appropriate, depending on the runoff chemical constituent(s), flow dynamics, soil conditions, position of wetland in the landscape, and other landscape characteristics. (From the United States Hydrologic Units [8-digit HUCs] and the United States Coastal Change Analysis Program [CCAP].)

plants and animals that require fluctuating wetland conditions and isolated upland areas. Thus, high perforation values may be considered as detrimental for some ecological functions and species and advantageous for others.

Fragmentation of wetlands may lead to increased inter-wetland distances because of the increases in the incidence and extent of other land cover types developing in the intervening spaces (e.g., farmland or human habitations). Accordingly, mean distance to closest like-type wetland (Figure 4.19) is an important metric to analyze in the project area because it may indicate the likelihood of nearby similar wetland habitat (e.g., neighboring emergent–emergent wetlands for migratory bird resting and foraging or neighboring forest–forest wetlands for migratory songbird resting and

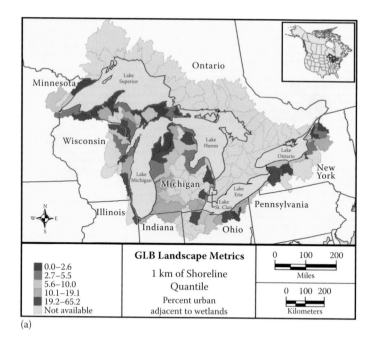

(a)

FIGURE 4.16

Percent urban adjacent to wetlands within (a) the 1-kilometer coastal region. The effect of runoff from areas adjacent on wetlands may be strongly influenced by land cover type immediately adjacent to a wetland. Thus, the percentage of urban land cover within 30 meters of wetlands is calculated by summing the total number of urban land cover cells within a single cell buffer (30-meter) zone on the perimeter of wetland patches in the reporting unit and dividing by the area of the buffer plus the area of the buffered wetland. Cells outside buffer zones for wetlands within each reporting unit are ignored. Other buffer distances may be more appropriate, depending on the runoff chemical constituent(s), flow dynamics, soil conditions, position of wetland in the landscape, and other landscape characteristics. (From the United States Hydrologic Units [8-digit HUCs] and the United States Coastal Change Analysis Program [CCAP].)   *(Continued)*

foraging). The mean (for a reporting unit) minimum distance to closest wetland patch, for example, the distance from each wetland patch to its nearest neighboring wetland patch, should be measured from one patch edge to another patch edge, and may consist of multiple measures (e.g., mean of the three nearest patches). This metric is useful in determining relative wetland habitat suitability at scales that are ecologically significant for specific plant and animal taxa, and demonstrates the importance of establishing the ecological endpoint(s) of interest prior to full development of this indicator.

As discussed briefly in earlier chapters, the Shannon–Wiener Index and Simpson's Index are two different ways of measuring the diversity and distribution of land cover types within a specific area of the landscape. The

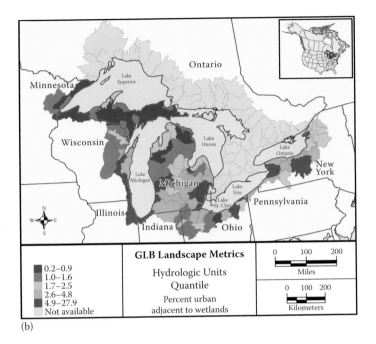

(b)

FIGURE 4.16 (CONTINUED)
Percent urban adjacent to wetlands within (b) the full HUC of the U.S. portion of the Great Lakes Basin. The effect of runoff from areas adjacent on wetlands may be strongly influenced by land cover type immediately adjacent to a wetland. Thus, the percentage of urban land cover within 30 meters of wetlands is calculated by summing the total number of urban land cover cells within a single cell buffer (30-meter) zone on the perimeter of wetland patches in the reporting unit and dividing by the area of the buffer plus the area of the buffered wetland. Cells outside buffer zones for wetlands within each reporting unit are ignored. Other buffer distances may be more appropriate, depending on the runoff chemical constituent(s), flow dynamics, soil conditions, position of wetland in the landscape, and other landscape characteristics. (From the United States Hydrologic Units [8-digit HUCs] and the United States Coastal Change Analysis Program [CCAP].)

Shannon–Wiener Index of land cover type diversity (Figure 4.20) is calculated as:

$$H = -\sum_{i=1}^{m} P_i * \ln P_i, \text{ where } P_i = \text{the proportion of land cover type } i. \quad (4.1)$$

Shannon–Wiener Index values increase as the number of land cover types within the reporting unit increases, with higher index value coastal areas having more diverse land cover (i.e., more diversity) than areas with lower index values. Because higher Shannon–Wiener diversity in areas does not necessarily indicate greater opportunities for a variety of elements (i.e., land

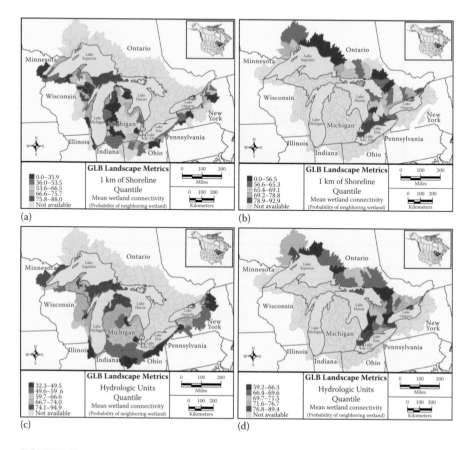

FIGURE 4.17
Mean wetland connectivity (probability) within/among the 1-kilometer coastal regions of the
(a) U.S. and (b) Canada watersheds of the Great Lakes Basin, and within/among the full watersheds of the (c) U.S. and (d) Canada portions of the Great Lakes Basin. Probability of a wetland
cell having a neighboring wetland cell is calculated using a moving 270-meter-square window
(9 pixels × 9 pixels) across the land cover. The boundaries between all pixel pairs where at least
one pixel is wetland are examined in the window. The metric is the number of boundaries
where both pixels are wetland, divided by the total number of wetland boundaries (regardless of neighbor land cover type). This metric gives a measure of how well the wetland is
connected within the window sample area, with high values being better connected than low
values. (From the United States Hydrologic Units [8-digit HUCs], United States Coastal Change
Analysis Program [CCAP], Canada Hydrologic Units [Subsubdivisions], and Ontario Ministry
of Natural Resources Data Set.)

cover diversity in this case includes agriculture and urban as well as natural land cover types), Simpson's Index (Figure 4.21) can be used to better
describe the distribution of the land cover in a region. Simpson's Index is a
quantitative measure of the evenness of the distribution of land cover classes
and is most sensitive to the presence of common land cover types within a

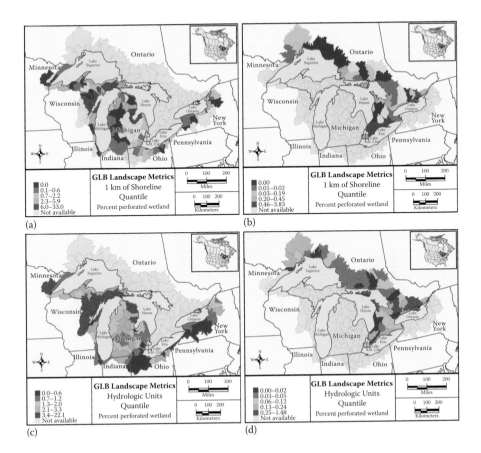

FIGURE 4.18

Percent perforated wetland within/among the 1-kilometer coastal regions of the (a) U.S. and (b) Canada watersheds of the Great Lakes Basin, and within/among the full watersheds of the (c) U.S. and (d) Canada portions of the Great Lakes Basin. Percentage of perforated wetland is calculated using a moving 270-meter-square window (9 pixels × 9 pixels) across the land cover. When the percent wetland in the window is greater than 60%, and greater than the window's mean wetland connectivity value, the wetland cell in the center of the window is categorized as perforated. The number of perforated wetland cells in the reporting unit is then divided by the reporting unit's total land area (i.e., the total number of cells in the reporting unit boundary minus those cells classified as water) to derive the percentage of perforated wetland. Perforated wetland generally consists of a patch of wetland with a center upland area, such as would occur if a small clearing were made within a patch of wetland, or if an area of wetland contained an interior upland. Perforated wetlands may be fragmented in this fashion to such an extent that do not provide a suitable interior habitat for some wetland species. However, the interspersion of upland and wetland conditions in perforated wetlands may provide a suitable habitat for some specialized plants and animals that require fluctuating wetland conditions and isolated upland areas. (From the United States Hydrologic Units [8-digit HUCs], United States Coastal Change Analysis Program [CCAP], Canada Hydrologic Units [Subsubdivisions], and Ontario Ministry of Natural Resources Data Set.)

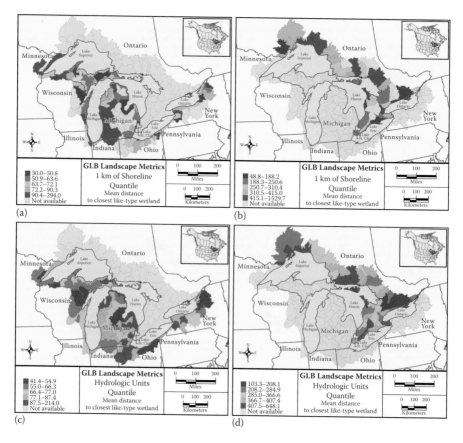

FIGURE 4.19

Mean distance to closest like-type wetland within/among the 1-kilometer coastal regions of the (a) U.S. and (b) Canada watersheds of the Great Lakes Basin, and within/among the full watersheds of the (c) U.S. and (d) Canada portions of the Great Lakes Basin. The mean minimum distance to the closest wetland patch is the average distance from each wetland patch to its nearest neighboring wetland patch in the reporting unit. Distances are measured from edge to edge and are reported in meters. This metric is useful in determining relative wetland habitat suitability, at scales that are ecologically meaningful for specific plant and animal taxa. (From the United States Hydrologic Units [8-digit HUCs], United States Coastal Change Analysis Program [CCAP], Canada Hydrologic Units [Subsubdivisions], and Ontario Ministry of Natural Resources Data Set.)

reporting unit. Simpson's Index values range from 0 to 1, with 1 representing perfect evenness of all land cover types within a reporting unit. The Simpson's Index is calculated as:

$$H = 1 - \sum_{i=1}^{m} Pi^2 \text{, where } Pi = \text{the proportion of land cover type } i. \quad (4.2)$$

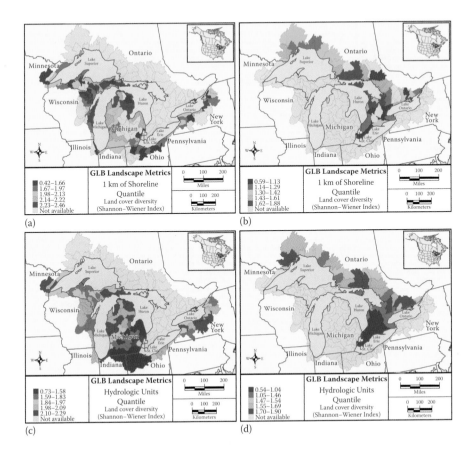

FIGURE 4.20
Land cover diversity (Shannon–Wiener Index) within/among the 1-kilometer coastal regions of the (a) U.S. and (b) Canada watersheds of the Great Lakes Basin, and within/among the full watersheds of the (c) U.S. and (d) Canada portions of the Great Lakes Basin. The Shannon–Wiener Index is one of several ways to measure the diversity of land cover types within a specific area of the landscape. The Shannon–Wiener Index is a measure of the diversity of land cover types within a reporting unit, and the index value increases as the number of land cover types within the reporting unit increases. A single land cover type results in a Shannon–Wiener Index value of zero. (From the United States Hydrologic Units [8-digit HUCs], United States Coastal Change Analysis Program [CCAP], Canada Hydrologic Units [Subsubdivisions], and Ontario Ministry of Natural Resources Data Set.)

4.1.4 Inferring Ecosystem Functions from Landscape Metrics

4.1.4.1 Water Quality Metrics Related to Wetlands of the Great Lakes Basin

As has been discussed, wetlands play an integral role in the environment, and they certainly play an important role in the hydrologic cycle of the Great Lakes Basin. They also provide important ecosystem functions and services that include flood storage during periods of high water and can act to improve

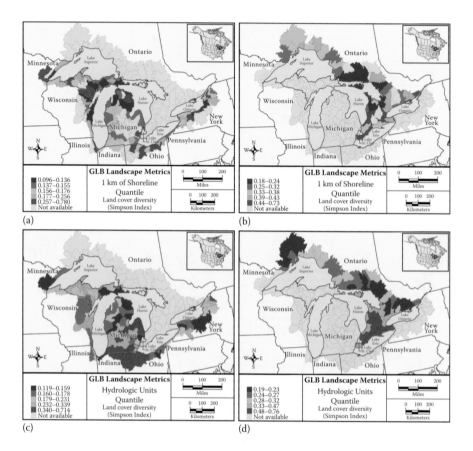

FIGURE 4.21
Land cover diversity (Simpson's Index) within/among the 1-kilometer coastal regions of the
(a) U.S. and (b) Canada watersheds of the Great Lakes Basin, and within/among the full watersheds of the (c) U.S. and (d) Canada portions of the Great Lakes Basin. Simpson's Index is one
of several ways to measure the diversity of land cover types within a specific area of the landscape. Simpson's Index is a measure of the evenness of the distribution of land cover classes
and is most sensitive to the presence of common land cover types within a reporting unit.
Simpson's Index values range from 0 to 1, with 1 representing perfect evenness of all land cover
types within a reporting unit. (From the United States Hydrologic Units [8-digit HUCs], United
States Coastal Change Analysis Program [CCAP], Canada Hydrologic Units [Subsubdivisions],
and Ontario Ministry of Natural Resources Data Set.)

the quality and safety of water resources in the Great Lakes. Wetlands in the
Great Lakes Watershed can cleanse surface and ground water before it enters
the shore waters (Lake Huron Center 2000) by intercepting, accumulating,
and transforming contaminants that are contained within soil particles that
travel in runoff from upland areas toward the open water areas. An increase
in soil erosivity (Figure 4.22) or erodibility (Figure 4.23) may indicate an
increase in the amount of the runoff to streams, lakes, and wetlands that

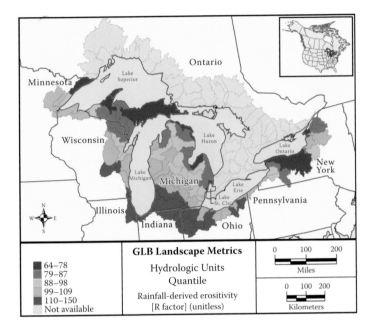

FIGURE 4.22
Rainfall-derived erosivity within the full HUC, among the watersheds of the U.S. portion of the Great Lakes Basin. This metric is a Revised Universal Soil Loss Equation (RUSLE) weighted-average rainfall-derived erosivity metric, which is derived from a Parameter-elevation Regressions on Independent Slopes Model (PRISM) 2-km grid, is computed on a cell-by-cell area basis. An increase in soil erosivity may indicate an increase in the amount of runoff of sediment and chemical constituents associated with sediment (e.g., phosphorus) to streams and lakes. (From the United States Hydrologic Units [8-digit HUCs], United States Coastal Change Analysis Program [CCAP], United States National Elevation Dataset [NED], and United States STATSGO Soils.)

may contain this eroded sediment and the chemical constituents associated with that sediment (e.g., phosphorus, road salts, and trace metals). Excessive amounts of this sediment, nutrients, or other chemical constituents within runoff may degrade surface water, ground water, wetlands, and the open water areas of the Great Lakes in general. Some watersheds in the Great Lakes have less sediment runoff than others, and thus are considered to be relatively "least impacted" by erosion and runoff.

An increase in surface roughness (Figure 4.24), a function of land cover and soil physical characteristics, may indicate a decrease in the amount of runoff and concomitant chemical constituents to streams, lakes, and wetlands. In addition to soil and general land cover characteristics, the presence of wetlands (by virtue of specialized vegetation and highly organic and clay soils) can have a tremendous influence on the reduction of sediment runoff to the open water of the Great Lakes. Wetlands slow down the movement of sediment and thereby trap pollutants in wetland vegetation's tissues. Thus, chemicals like nitrogen and phosphorous (commonly associated

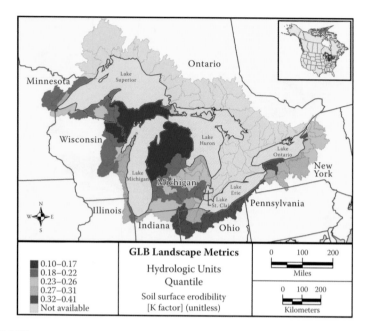

FIGURE 4.23

Soil surface erodibility (K factor) within the full HUC, among the watersheds of the U.S. portion of the Great Lakes Basin. This metric is a RUSLE weighted-average effect of inherent soil surface erodibility (K factor), which is from STATSGO data, is computed on a cell-by-cell area basis. An increase in soil erodibility may indicate an increase in the amount of runoff of sediment and chemical constituents associated with sediment (e.g., phosphorus) to streams and lakes. (From the United States Hydrologic Units [8-digit HUCs], United States Coastal Change Analysis Program [CCAP], United States National Elevation Dataset [NED], and United States STATSGO Soils.)

with agricultural runoff) and pesticides are taken up by the root systems of wetland vegetation (National Wetland Plants List 2012), which incorporates them into plant tissue (Lopez 1997), subsequently incorporating these constituents into the organic and clay soils, potentially for very long periods of time (Environment Canada 1995). In areas where surface roughness is low, this surface water cleansing process that wetlands can provide may be critical in preventing eutrophication, which is a major human health and nuisance issue, as well as a threat to aquatic plants and animal species.

The Great Lakes case study provides a clear example of how a successful project is conceptualized and planned, and can serve as a template for other broad-scale projects, including the critical need to be aware of, and incorporate the complexity of, gradients across multiple regions of a vast landscape. The specific parameters analyzed in this case study can be replaced by any number of similar/analogous parameters of other projects across the world to meet their specific goals. It is important to recognize that, no matter the geography of interest to you as the reader (whether the next watershed over

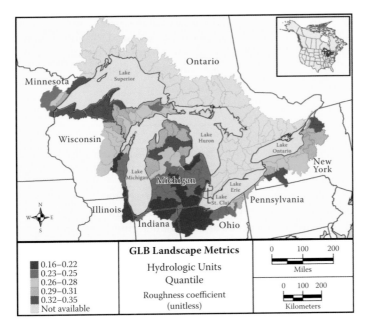

FIGURE 4.24
Roughness coefficient (unitless) in the 1-kilometer coastal region and within the full HUC, among the watersheds of the U.S. portion of the Great Lakes Basin. This metric is a SEDMOD weighted-average Manning's n surface roughness coefficient. An increase in surface roughness may indicate a decrease the amount of runoff to streams and lakes, which may contain sediment, road salts, or other compounds. (From the United States Hydrologic Units [8-digit HUCs], United States Coastal Change Analysis Program [CCAP], United States National Elevation Dataset [NED], and United States STATSGO Soils.)

from the Laurentian Great Lakes, within the African Great Lakes Basin, or any other set of watersheds around the world) a focused analysis of parameters within the study area is always necessary, specifically for scientific, societal, and policy importance and relevance, and such analyses should, importantly, be highly dependent upon the focal biophysical information that is available or attainable, as well as the needs of decision makers. Balancing these two elements, information availability and the applicability of that information to decision makers' needs, is at the crux of all of the case studies in this book. The practicalities of balancing these two critical elements is particularly necessary in the diverse and vast landscapes of the Great Lakes, which certainly provide an ideal opportunity to demonstrate a number of gradients simultaneously, and to also reveal the utility for a synoptic view of such a vast landscape, and how that balanced approach best serves the needs of the populace and societies of the physical environment under assessment.

In the next section of this chapter, an analysis of coastal regions of California and North Carolina utilizes landscape ecological techniques to tackle the broad-scale impacts of potential sea level rise, at a global scale.

Accordingly, the next case study is intended to soundly demonstrate how the complexity of various practical factors of cost, and validity of assessments, can be successfully balanced, particularly across extremely vast areas of the planet and in the face of global scale drivers of change.

4.2 Case Study: Impacts of Sea Level Rise on California and North Carolina Coastal Ecosystems

Extreme change in tidal wetlands as a result of sea level rise in the twenty-first century, globally, is no longer debated. Therefore, the extent and the degree of the loss of coastal land is a critical and challenging question for geographers, ecologists, and policy makers to address now and in the coming decades. Coastal ecosystems such as marshes and lagoons are among the most susceptible ecosystems to sea level rise (SLR) because of their location on the front lines of coastal areas, as well as the short-term and long-term effects of changes in inundation, salinity, and associated soil characteristics. The impact of changes in sea level rise are pressing and immediate, especially for a number of communities around the world such as coastal indigenous communities of Alaska, the low-lying atolls of the Western Pacific Islands, and the Outer Banks of North Carolina, to name just a few, where residents are currently experiencing the physical changes in local coastal landscape configuration and conditions.

Tools for modeling such a vast and complex set of landscape changes are limited by the amount of synoptic data that can be brought to bear on the coastal regions affected by SLR, and thus remote sensing data have been very beneficial in enabling the development of new and illustrative SLR models in these coastal areas (Klemas 2009) over the past decade. Among the increasingly utilized modeling tools to predict and measure change in tidally influenced wetlands is the Sea Level Affecting Marshes Model (SLAMM), which is most applicable at local to regional scales. The development of SLAMM by Dr. Richard Park was originally funded by the Environmental Protection Agency in 1985, and its success as an effective predictive tool is in part a result of the increased availability of large landscape data sets over the past 30 years, and increased computational power (i.e., speed at lower costs) as well. SLAMM has gone through several revisions since 1985, and the version utilized for the models in this case study is SLAMM 6.0.1, which was originally made available in 2010. The U.S. Environmental Protection Agency, the U.S. Fish and Wildlife Service, the Department of Defense, San Diego State University, the Nature Conservancy, and the Gulf Coast Prairie Land Conservation Cooperative are some of the many organizations that are currently utilizing SLAMM for land use planning and restoration of ecosystems in selected coastal areas. Another approach to assessing the coastal impacts of SLR, which is much more practical for applications across larger landscape areas (i.e., long lengths of coastal areas and for inland areas along those coast-lines, such as the entire coast of the eastern United States), is based upon ele-vational changes along those coastlines, and is less computationally complex and intensive than the SLAMM. This less complex and intensive approach is sometimes referred to as a "bathtub model," where modeled SLR is based upon water levels rising uniformly and inundating coastal land based solely upon the topographic relief of a coastal area. Because the "bathtub model" method is less computationally complex and intensive than the SLAMM, it allows for broader-extent application to coastal regions at reduced effort and cost. Cooper et al. (2008) specifically used this simpler bathtub model, or Elevational Sea Level Model (ESLM) method, that is, using solely future sea level rise predictions and coastal topographic elevation. The primary input layer utilized in the ESLM method in this case study is a digital elevation (bare earth terrain) model, adjusted to the local sea level in a GIS and using the same rising sea level predictions used by most predictive impact models, including the SLAMM, that is, the SLR predictions of the Intergovernmental Panel on Climate Change (IPCC, https://www.ipcc.ch/, checked January 25, 2017). Accordingly, this case study provides an outlook for the future coastal landscape, given rising sea level predictions of the IPCC, based on two dif-ferent assessment methodologies, the SLAMM and the ESLM.

Because coastal marshes are among the most susceptible ecosystems to climate change, especially accelerated SLR, the use of one or both the ESLM and the SLAMM may be critical to understanding better how projected coastal land cover conversion will progress over the next years, decades, and century, in the face of climate change. The use of remote sensing data inputs for predicting those changes can create outputs that are based on both the geometric and inferred relationships among coastal land cover and land use classes. The SLAMM utilizes a number of assumptions about the interaction of those mapped land cover classes and hydrologic input variables, which are outlined in general in this chapter for two specific geographies on both the east and the west coasts of the United States, specifically (1) the south-western U.S. Pacific Coast (Carpinteria Marsh, California) and (2) the U.S. Atlantic Coast (Harkers Island Wetland Complex, North Carolina).

This case study is intended to provide the reader with a clear understand-ing of two applications of remote sensing data for two sea level modeling approaches, SLAMM and ESLM. The comparison of outputs from both the SLAMM and the (computationally simpler) ESLM approach, within two very different coastal environments of the United States, provides potential appli-cations to a diversity of coastal areas around the world. The SLAMM simulates dominant processes involved in land cover conversions and shoreline modi-fications during long-term sea level rise; however, broad-scale application of the SLAMM model is challenging and likely not practical because of the lack of sufficient, uniform, and consistent data for input to the computationally intensive model, especially along vast lengths of coastline, such as nationally or continentally. The computationally less intensive ESLM method provides analogous results as SLAMM (however, quite different magnitudes of change), which might be useful in understanding the applicability of either or both methods, perhaps for augmenting the finer-resolution and more parameterized SLAMM outputs in certain coastal areas of interest that lend themselves to the relatively more generalized outputs from the ESLM. Data inputs for the SLAMM include SLR projections, as well as accretion, overwash, saturation, and sedimentation rates, as available. Alternatively, the ESLM is based solely on land elevation and sea surface (level) height projections, without accounting for any other factors, such as accretion or sedimentation. It should be noted that the lesser requirement for data input to the ESLM is advantageous in terms of computational simplicity, and thus likely reduced time and cost for applying this method. However, there is a concomitant reduction in the detail and complexity of outputs from the ESLM model. It should be understood that this is a fundamental tradeoff in the use of remote sensing for landscape ecology, that is, between the depth and richness of information outputs and cost. It is recommended that the correct balance of these two factors (i.e., the amount and specificity of information, and cost) for any project be decided in advance in collaboration with team members, particularly the decision makers.

4.2.1 California and North Carolina Coastal Landscapes: Context and Background

Carpinteria Marsh, California, was selected for this case study because it is an example of a wetland complex, embedded within a complex landscape matrix, that is highly likely to be affected by SLR along the western coast of the Americas, and primarily because of its typical characteristics as a marsh and physical location on that coast that is, with little to no opportunity for upland migration of wetland hydrologic conditions, a result of the fully developed land currently along its upland perimeter. Marshes such as this one are relatively rare on the west coast of the United States, among a handful of remaining intact estuaries of the California coastal region. Carpinteria Marsh is also home to many rare and endangered birds, such as the light-footed clapper rail and Belding's savannah sparrow; is an important west coast fish nursery for halibut and other marine and estuarine fish; and serves as a physically protected area for fish and other organisms to seek refuge within. Carpinteria Marsh is also prized by local and regional members of the community, and utilized for environmental education and scientific research by many individuals living in those communities, as well as visitors. Situated in southern Santa Barbara County, California, and embedded within the municipality of Carpinteria (population 13,040 [U.S. Census 2010]), this 230-ha marsh is a remnant of a once much larger wetland complex that filled the Carpinteria Valley, now converted to agricultural and residential land. Avocados, exotic fruits, and flowers are the major crops in the local area (Figure 4.25). The marsh area is owned and managed by the city of Carpinteria and the University of California, Santa Barbara. Carpinteria enjoys a mild Mediterranean climate,

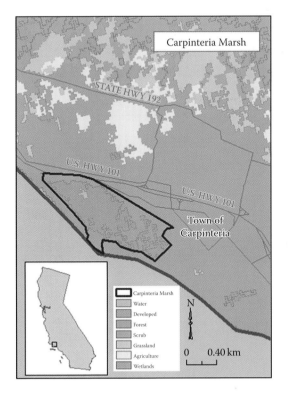

FIGURE 4.25
Locational map of Carpinteria Marsh, California, including land cover and land use in the vicinity.

with a wet season from December through March and an average rainfall of 457 mm. Pacific storms, especially during El Niño years, can bring heavy surf, wind, and devastation along this coastline. The fact that Carpinteria Marsh is among just a handful of marshes that remain along the western coast of California makes it critically important to better understand how this and the other last remaining marshes of California may change in size and ecological functions over the coming years from predicted continued SLR.

Harkers Island coastal wetland complex (Figure 4.26) is in North Carolina and encompasses the small towns of Beaufort (population 4,039 [U.S. Census 2010]) and Morehead City (population 8,661 [U.S. Census 2010]), as well as the Cape Lookout National Seashore. Population of Harkers Island itself is 1,207 (U.S. Census 2010) and it is an unincorporated area of Carteret County, North Carolina. A large number of the local and regional fishery species depend directly upon coastal marshes of the region for some portion of their life cycle (Voss et al. 2013; Sheaves et al. 2015). Thus, the people of this region, who rely heavily on tourism and commercial fishing as major sources of income for residents, directly rely on coastal wetlands of the study area to support the ecosystem services that maintain fishing and diving industries, including sport

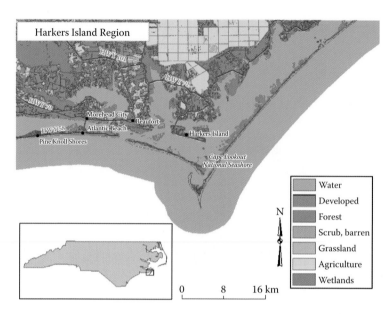

FIGURE 4.26
Locational map of Harkers Island region, North Carolina, including land cover and land use in the vicinity.

and recreational fishing, as well as commercial fisheries that employ many residents. The case study region of Harkers Island is approximately 125,000 ha with low topographic relief throughout, comprised of numerous and extensive forested, shrubby, and herbaceous wetlands across the extensive coastal plain, which extends much further inland than similar areas on the west coast of the United States. Due to the location, and regular weather patterns in the region that bring extreme storms, the study region is highly susceptible to hurricanes and storm surge, as was memorably demonstrated during many hurricanes over the last century, including several most notably: Hurricane Connie (1955), Hurricane Donna (1960), Hurricane Floyd (1999), and Hurricane Irene (2011). Such storms bring with them, too, cumulative costs to society, including infrastructural costs in the many hundreds of billions of dollars.

4.2.2 Determining Terrestrial Conditions and Changes in Sea Level

4.2.2.1 SLAMM Approach

4.2.2.1.1 Carpinteria Marsh

SLAMM simulations required several GIS data layers, each prepared in ArcGIS 9.3 (ESRI, Redlands, California): high-resolution LiDAR (Light Detection and Ranging) data, which reduced model uncertainty (Gesch 2009); a USGS 3-meter DEM (digital elevation model) for Southern California (Barnard and Hoover 2010), derived from 10-meter LiDAR (extending approximately 500 meters inland); a tsunami DEM (NOAA); and 5-meter ifSAR (interferometric synthetic aperture RADAR) data (Gesch 2009). The LiDAR data (2005) were used for the DEM input into the model and all data were projected to NAD 83 Zone 11, N UTM producing a slope layer.

National Wetlands Inventory (NWI) 1:24,000 quadrangle data were subsetted to cover solely the study area, to minimize computational requirements. SLAMM wetland codes used were based on NWI data and found in the SLAMM technical documentation (Clough et al. 2010) and within database files provided with the model software. These codes were added to the NWI file prior to converting it to raster data, at the same resolution as the DEM.

National Land Cover Database (2001) impervious surface data were used to produce a developed/undeveloped land layer in SLAMM. Developed land was set to "protected" in the model due to the high likelihood that the city of Carpinteria would take infrastructure measures in the future to counter sea level rise impacts on developed areas. The data were reclassified to greater than 25% impervious for developed dry land per the SLAMM User's Manual (Clough and Larson, 2010). After all of the preceding steps, the impervious and SLAMM code layers were then combined (Figure 4.27).

Site parameters included the NWI photo data (2006), which were the initial data for the SLAMM simulation. The onshore direction of water flow was set to south in order to match the closest cardinal direction category offered in the model; the flow direction was used to determine if a land raster cell is adjacent to open ocean, which determines whether inundated dry land converts to beach or wetlands in simulations of future conditions (Clough and Larson 2010). The "A1B" mean IPCC sea level rise rate was utilized, which assumes rapid economic growth, global population that peaks around 2050 and then declines, and the rapid introduction of new technologies (Houghton et al. 2001).

Historical sea level rise trends of 2.5 mm/yr were utilized to estimate subsidence in the study region. Sea level rise trends were interpolated based on the nearest National Oceanic and Atmospheric Administration (NOAA) tide stations: in Santa Barbara (13 kilometers west) the historic trend was 1.25 mm/yr; on Rincon Island (8 kilometers southeast), the historic trend was 3.22 mm/yr, using mean sea level (MSL) from the NOAA (2010) Tides and Currents Co-Ops (https://tidesandcurrents.noaa.gov, checked January 25, 2017). Inverse distance weighting was applied to MSL in ArcGIS. Accordingly, calculated historical sea level rise trends on the west and east side of Carpinteria Marsh was 2.36 mm/yr and 2.63 mm/yr, respectively, with an average of 2.5 mm/yr at the marsh.

The North American vertical datum from 1988 (NAVD 88) was subtracted from the mean tide level (MTL) to adjust the DEM to zero, and the datum was then utilized in the SLAMM (Clough and Larson 2010). Datum calculation (from NOAA tides and currents datum sheets) differences for Santa Barbara and Rincon Island were 0.837 mm and 0.838 mm, respectively. Tidal range for Carpinteria Marsh, utilized for the SLAMM, was 1.01 m, which was recorded from a pressure sensor in the marsh (Sadro et al. 2007). Salt

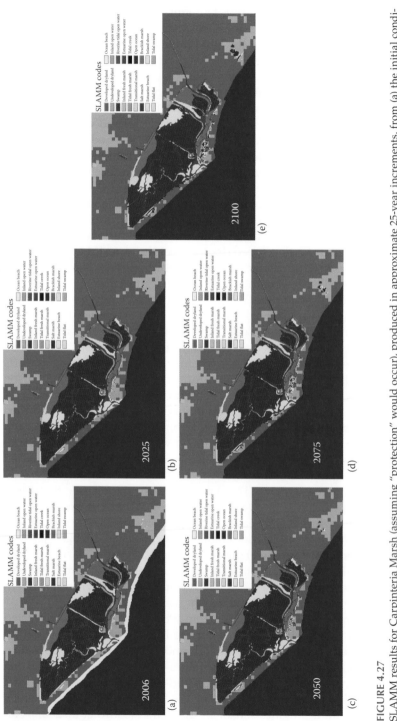

FIGURE 4.27
SLAMM results for Carpinteria Marsh (assuming "protection" would occur), produced in approximate 25-year increments, from (a) the initial condition in 2006, to (b) 2025, to (c) 2050, to (d) 2075, to (e) 2100. The SLAMM methodology allows for a visualization of land cover change at a frequency that enables a better understanding of the potential impacts of sea level rise on coastal zones. The frequency of 25 years optimizes the processing time and effort for this relatively computationally intensive approach.

elevation (mean high water spring), or meters above MTL, was 1.37 m for Santa Barbara (no data for Rincon Island) as recorded in historical NOAA tides and currents datasheets. Maximum monthly mean tide level was 1.37 meters; the MTL (1.83 m) was subtracted from the maximum monthly MTL (3.20 m).

Erosion rates used were those provided as SLAMM defaults. Marsh erosion was set to 2 horizontal m/yr; swamp erosion of 1 m/yr and tidal flat erosion of 0.5 m/yr were utilized (per Clough, et al. 2010). Revell (2007) recorded 2 m/yr erosion at the inlet to Carpinteria Marsh. Local accretion data were not available for Carpinteria Marsh; however, 320 km south along the California coast, Tijuana Slough, with similar climate and physiography, had an accretion rate of 7–12 mm/yr (Weis 2001); SLAMM technical documentation suggests accretion rates of up to 5 mm/yr (Clough et al. 2010). We utilized the available information to parameterize overall vertical accretion at 5 mm/yr, brackish vertical accretion at 4.75 mm/yr, and salt marsh vertical accretion at 4.6 mm/yr. Sensitivity analyses of various accretion rates were run, with negligible changes in "estuarine creep" toward the upland areas by 2100, including after a 7 mm/yr input into one of the simulations. A 5 mm/yr beach/tidal flat sedimentation rate was utilized, based on Mission Bay (San Diego County, California), the closest similar slough area with sedimentation data, approximately 195 km to the south (Mudie 1980). For the reasons mentioned earlier, developed land in the Carpinteria area surrounding the study site was set to "protected" in the SLAMM.

SLAMM results can be produced in time steps as short as one year, however we chose to output results using approximate 25-year increments, as suggested in the SLAMM technical documentation (Figure 4.27), which reduces the computational requirements of data processing in the model, which is particularly important if the SLAMM were used across vaster areas of coastline. To fully illustrate the total predicted impact of coastal ecosystems, and to demonstrate the power and promise of utilizing remote sensing approaches for landscape ecology, comparisons of potential protection, or non-protection actions on the part of decision makers in the region were used. Comparing the non-protected adaptation strategy (Figure 4.28) to the protected strategy (Figure 4.27) at Carpinteria Marsh, the largest changes for the twenty-first century of predicted SLR are modeled to be in those areas of developed dry land, predicted to be reduced by 5.97 ha, and open ocean areas that are predicted to be reduced by 3.6 ha, in the non-protected scenario. Results from both the protected and non-protected SLAMM runs show ocean beach will decrease the most (approximately 22 ha), as open ocean increases (by approximately 30 ha). A detailed summary of inferred changes in land cover is included in Section 4.2.3.

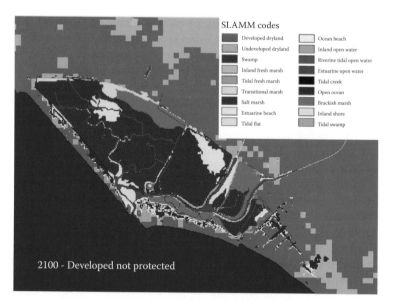

FIGURE 4.28
Non-protected condition prediction for Carpinteria Marsh in 2100, utilizing SLAMM.

4.2.2.1.2 Harkers Island

SLAMM was run on a 38-km × 64-km area of Harkers Island, which is a substantially larger area than Carpinteria Marsh. The region was then partitioned into three sub-regions to accommodate the limited memory capability of SLAMM software, given the relatively large geographic area under analysis. Several National Wetlands Inventory (NWI) quadrangles (1982, the most recent year) were downloaded, spatially merged, and assigned SLAMM codes. A data field for dikes was added to include the dikes at the study site, which had not been previously part of the NWI data. The NWI data were converted to raster, based on the digital elevation model (DEM) cell size of 6 meters. The DEM (2006) was obtained from NOAA (National Oceanic and Atmospheric Administration), and a slope file was created from that DEM. The impervious layer was also added to the Carpinteria Marsh analysis process, and the flow direction offshore was set to south for the west and middle regions, and east for the eastern region of the Harkers Island study site.

The historical sea level trend (NOAA tides and currents datum sheets) was 2.57 mm/year at the only local recording station, which is in Beaufort, North Carolina; the NAVD 88 correction was also obtained from NOAA datum sheets. The tidal range was averaged for the various stations in each region and the salt elevation (mean high water spring), or meters above mean tide level (MTL), was averaged for each region based on historical NOAA tides and currents data stations across the study region. Marsh and swamp erosion rates were set to 2 and 1 horizontal meters/year, respectively, from both a Georgia and Chesapeake Bay study (Ehman 2008; Glick 2008). The tidal flat erosion rate of 6 horizontal meters/year was taken from the Chesapeake Bay study that was based on simulations in GA, SC, and MD (Glick 2008). Both Glick and Ehman used the same erosion rates, and therefore these rates were used since North Carolina lies between the Chesapeake Bay and South Carolina. As is often the case in coastal areas around the world, local field-based data were not available in the case study area of North Carolina. Salt marsh accretion was averaged to 3 mm/year based on the rates from Benniger (1985) (2–4 mm/year), Moorhead and Brinson (1995) (2 mm/year), and Cahoon et al. (1999) (3.7 mm/year). Brackish vertical accretion input was 4.3 mm/year, and tidal fresh vertical accretion was 4.8 mm/year (Ehman 2008; Glick 2008). A beach tidal flat sedimentation rate of 0.5 mm/year was selected, based on Glick (2008). Frequency of large storms (overwash) was set to 25 years, the default. Protected developed land was selected and the IPCC A1B mean was the climate scenario entered into the model, as was the case in Carpinteria Marsh. Note that parameter data for Carpinteria Marsh was restricted to a much smaller geographic region and could thus be refined much better than the relatively larger Harkers Island area. Also, note that tidal measurements, sedimentation rates, and accretion rates are easier to determine for a confined wetland such as Carpinteria, as compared to the North Carolina coast, which is quite varied and heterogeneous throughout, with barrier islands, estuaries, and numerous other types of wetlands present, such as forested (swamp) wetlands. SLAMM results through the year 2100 in the Harkers Island project area are depicted in Figure 4.29.

4.2.2.1.3 SLAMM Summary

The summary of SLAMM data input, both derived from remote sensing data and field-based data at both Carpinteria and Harkers Island project areas, is as follows:

1. Elevation data based on LiDAR, which are the preferred model input, due to these data being of high spatial resolution and precision. Consequently, the input layer to SLAMM is a 3-meter spatial resolution LiDAR-based digital elevation model. The slope layer was derived from the elevation data. Note that SLAMM calculates inundation based on the minimum elevation and slope of each cell.

2. Local erosion and accretion rates: Erosion is triggered in the model given a maximum fetch threshold and proximity of the marsh to estuarine water or open ocean. SLAMM erosion defaults were selected for the runs. Saturation is the response of the water table to rising sea level as coastal wetlands migrate upland. Accretion is the vertical rise of the marsh due to the buildup of organic and inorganic matter. Accretion data were estimated based on other Southern California marshes.

3. SLAMM wetland categories: National Wetlands Inventory data (U.S. Fish and Wildlife Service) were adjusted to the SLAMM codes from

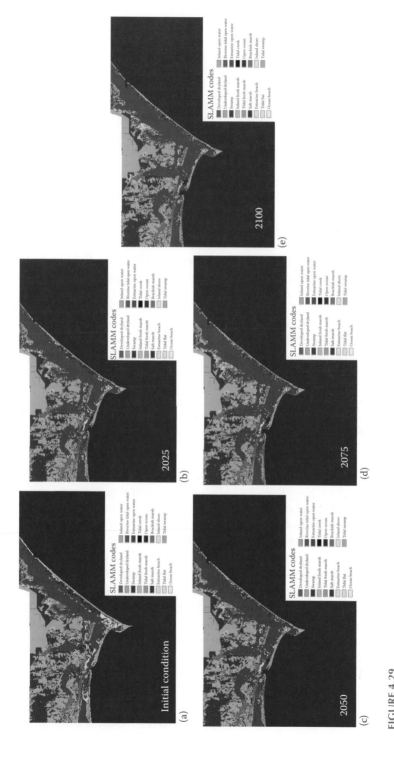

FIGURE 4.29
SLAMM results for the Harkers Island region (assuming "protection" would occur), produced in approximate 25-year increments, from (a) the initial condition in 2006, to (b) 2025, to (c) 2050, to (d) 2075, to (e) 2100.

the manual. The NLCD impervious layer was added to the NWI data and given a value of "developed" or "undeveloped" (upland) dry land.

4. NOAA station data: Local sea level trend data, tide range, and mean high water spring was input and adjusted to the local area; these data were corrected to the mean tide level datum used in SLAMM.

5. Sea level rise rate: The A1B mean Intergovernmental Panel on Climate Change sea level rise rate (IPCC 2001) was selected as it is an average sea level rise scenario, commonly used in research, and is hardwired into the SLAMM model. The rate is 0.39 meters by 2100, which is a conservative estimate. More recent rate projections, such as from Rahmstorf (2007), are greater, which assumes more rapid economic growth, a global population that peaks mid-century then declines, and the rapid introduction of new technologies.

4.2.2.2 ESLM "Bathtub Model" Approach

4.2.2.2.1 Carpinteria Marsh

ArcGIS was utilized to produce and compare the predictions from the ESLM "Bathtub Model" with those from the SLAMM outputs. To that end, the IPCC 2001 mean for A1B was used as input (0.39 meters by 2100) into the ESLM, as it was for the SLAMM approach. The process was run on Rahmstorf's (2007) update of the IPCC figure maximum of 1.4 meters, per Cooper et al. (2008). Rahmstorf included ice sheet and ocean heat intake models that were not included in the IPCC calculations. Rahmstorf has since then updated the figure to 1.9 meters by 2100, so the actual changes in sea level may be even greater (Vermeer and Rahmstorf 2009).

NOAA historical trend data interpolated from nearby Rincon Island, California, and Santa Barbara, California, gauges were 2.5 mm/year. The global rate of 1.5 mm/year (Cayan et al. 2006) and the local 2.5 mm/year used to obtain a sea level rise rate of 1 mm/year for 2100 predictions and the resultant differential of 1 mm/year was added to the A1B mean and Rahmstorf (2007) maximum for the adjusted SLR rates of 0.49 and 1.5, respectively, based on Cooper et al. 2008. Data were then adjusted to the NAVD 88 datum, and 0.83 was added for new elevation values of 1.32 and 2.33, respectively. The LiDAR 3-meter DEM from Barnard and Hoover (2010) was reclassified into values of 0–1.32, 0.32–2.33, and all categories greater than 2.33. The SLAMM was then run, utilizing the assumption that land was unprotected from inundation, to match more closely with the ESLM processes.

As part of the SLAMM runs, the 100-year event scenario from Cooper et al. 2008 was utilized in Carpinteria Marsh; a 0.3-meter increase in SLR would shift the 100-year storm surge-induced flood event to once every 10 years (Cayan et al. 2006). The 100-year flood level base elevation (FEMA) was 3.66 meters for Carpinteria; these data were downloaded from the Pacific Institute (2008), then averaged for the marsh. The FEMA base elevation values were added to the adjusted IPCC mean and NAVD 88 adjustment: that is, 3.66 + 0.49 + 0.83 = 4.98 meters as the A1B mean, and 3.66 + 1.5 + 0.83 = 5.99 meters for the Rahmstorf maximum. SLAMM was then run with the same 100-year flood data and compared to the ESLM processes.

The ESLM, using the A1B mean sea level IPCC projection, predicts more open water than the SLAMM model for Carpinteria. Using the Rahmstorf projection, and much more inland open water is projected; the SLAMM output predicts the open water regions from the ESLM to be a mix of tidal flats, salt marshes, and open water. Comparing the 100-year event process with SLAMM unprotected results, the 100-year process shows most of the region covered in water for both the IPCC and Rahmstorf scenarios. SLAMM output in these regions is mostly tidal flats, open water, and inland shore.

4.2.2.2.2 Harkers Island

The same general modeling method used for Carpinteria, based on Cooper et al. 2008, was applied to Harkers Island, North Carolina. The SLR rate from NOAA for nearby Beaufort was 2.57 mm/year, which was higher than the 1.5 mm/year global mean average (Cooper et al. 2008). The global mean was subtracted from the local rate, for a rate of 1.07 mm/year that was added to the local NOAA adjustment (i.e., 0.107 m + 0.39 m [the A1B mean, 2100] = 0.497 m), and added to 1.4 (Rahmstorf 2007) = 1.51 m. The data were adjusted for the NAVD 88 correction by subtracting 0.12 (MTL-NAVD), equaling –0.38 m and –1.39 m for the A1B mean and Rahmstorf rates, respectively.

The 100-year base flood elevation at Harkers Island was 2 meters, from a Carteret County Flood Insurance Study (FEMA and the State of North Carolina 2003). The adjusted SLR value for the A1B scenario for 2100 in the previous step (0.38) was added to the 100-year base flood elevation of 2 meters, and was also added to Rahmstorf's total value of 1.39 for an overall total of 3.39 meters.

4.2.2.2.3 ESLM Summary

The summary of ESLM data input, both derived from remote sensing data, and field-based data at both Carpinteria and Harkers Island project areas is as follows:

1. Elevation data: 3-meter LiDAR layer
2. NOAA Station data: Historical trend data was used to adjust the local sea level to the A1B IPCC scenario and the Rahmstorf projections; the local adjusted rate was added to the A1B and Rahmstorf global projections and adjusted to the local NAVD88 datum

4.2.3 Inferring Ecosystem Conditions Using Associated Sea Level

Results of the SLAMM demonstrate landscape change (i.e., losses) in several land cover categories by 2100 for the Harkers Island region, with the largest losses in Undeveloped Dry Land (2,162 hectares), swamp (1,531 hectares), tidal flats (1,421 hectares), flooded marshes (751 hectares), and ocean beaches (568 hectares). Land cover categories that increased the most included transitional salt marsh (1,174 hectares), estuarine open water (4,177 hectares), and open ocean (1,356 hectares).

The ESLM, using the A1B mean sea level IPCC projection, shows more open water than the SLAMM model for the Harkers Island region. Using the Rahmstorf projection, much more inland open water is predicted. The SLAMM output shows the open water regions from the ESLM covered mostly with brackish marsh and tidal swamp; results are displayed for the simulated 100-year storm prediction. Even when SLAMM was set to "not protected" for developed land on Harkers Island, the ESLM approach predicts much more of the Harkers Island region becoming inundated in the future sea level scenario than the SLAMM approach.

For the A1B projection using the DEM, the SLAMM result at Carpinteria Marsh shows more inland open water than the ESLM; tidal flat, salt marsh, and open water category conversions are much more prominent for the SLAMM in 2100 than for the ESLM.

Overall, for both coastal case study areas, by comparing SLAMM results with the ESLM approach, it is apparent that the results are quite different. The ESLM approach indicates more SLR impacts in the coastal zone occurring in the future than the SLAMM model (Figure 4.30). SLAMM is a technically detailed program and has more input variables, which is time-consuming to learn, and the ESLM process is relatively quick and easy to apply. Neither, however, are trivial nor superior in their outputs, but rather

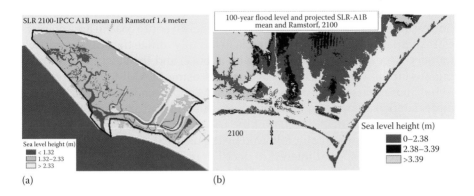

(a)   (b)

FIGURE 4.30
By comparing the Elevational Sea Level Model (ESLM) approach in (a) the Carpinteria Marsh and (b) the Harkers Island region to the SLAMM approach, it is apparent that the ESLM approach indicates more SLR impacts in the coastal zone occurring in the future than the SLAMM model, which is likely a function of the relative simplicity of the ESLM. Neither the ESLM nor the SLAMM method is trivial nor superior to one another in terms of their outputs, however, there is a need to balance "information needs" with overall project costs. The results shown are for the year 2100 at both study site locations.

are two different methods that utilize remote sensing and field-based data to arrive at similar (but not the same) results. Indications are, particularly considering the audiences involved and the scales and frequency of analyses needed, that a combination of the ESLM and the SLAMM processes might work the best, depending on the application, needs of decision makers, and cost sensitivities. The ESLM may be adequate for vast lengths of coastline and provide good general information, from which more detailed SLAMM analyses can emerge, particularly for wetland and estuarine areas that are heterogeneous or complex in topography and land use. However, SLAMM is much more field-data intensive, and dependent, and such data are often not available (note that we struggled to gather sufficient data for our SLAMM inputs at numerous junctures in the data preparation, described in detail previously), particularly data inputs like accretion and erosion rates, as well as the full mapping of dikes and other water infrastructure within and in the flow regime of particular wetland sites.

Incorporating periodic effects of El Niño, which is a very important set of environmental factors along the California coast, is also likely very important to include in order to improve precision and accuracy of both models. Other key questions that need to be answered are related to seasonal longshore flow of sand along the California coast, potentially impacting the results of SLR within these marshes up and down the coast. Nearshore sediment transport could also be added to improve the models, as well as cliff erosion rates, to improve predictions, if made available.

Most importantly, as related to the subject matter of this book, it is important to recognize that due to computational complexity, data unavailability, and capacity of computing, the SLAMM can only deal well with relatively small areas of the coastal zone dynamics at one time (i.e., multiple complex processing sessions may be necessary to model an entire coastal area of interest). Alternatively, it might be desirable to initially utilize the ESLM approach for very large sections of a coastline all at once, and then supplement these initial outputs with selected focal analyses using SLAMM, as desired. For initial, more coarse level, analyses, the ESLM approach works well, but for a more detailed analysis and for heterogeneous wetland areas, the SLAMM is likely most appropriate. The use of the ESLM for use along large regions of coast, perhaps continentally, followed by the use of SLAMM for high priority areas or areas of special interest to decision makers could optimize the use of high resolution remote sensing data, and allow for efficient reevaluation of these models as global climate predictions are adjusted, and land cover or land uses change, in the coming years and decades.

During the course of this project, several areas of improvement were discovered, which would improve the modeling of landscape condition, including the analysis of cliff erosion along the California coast, given that cliffs will erode on average of 54 meters by 2100 in the Santa Barbara region, at a minimum (Heberger 2009). On the eastern U.S. coast, hurricane-related surge data could be incorporated and improve North Carolina region predictions, as well as other similar areas along that coast. Current remote sensing imagery would also be very helpful in updating the (now 1982) NWI data for the North Carolina coastal area, in order to improve the accuracy of predictions for SLAMM and the ESLM, whichever is utilized for the areas for determining the impacts of sea level rise in the future.

The case studies in this chapter provide an improved understanding of the utilization of remote sensing to determine broad characteristics about vast areas of the landscape, such as was also the case in the entire Great Lakes Basin, with the leveraging of detailed field data to focus in on landscape change that is predicted to occur, as possible, given the limited ability to obtain field-based data across vast landscape areas, in this case along coastal landscapes. The use of field data in coordination with remote sensing, once again, redeems the promise of landscape ecology as a method for determining the factors that improve our understanding of ecosystem condition, as well as human health and well-being. The improvement of our understanding comes by way of specialized models of ecosystem condition, which can be applied among millions of hectares of wetlands, forest, and other ecosystems across vast landscapes that are changing all the time, and also by improving the prediction of ecosystem impacts from changes in the environment, such as sea level rise or changes in landscape configuration/use, over the coming years, across the globe.

The following chapter further focuses the promise of the use of remote sensing for landscape ecology, further enabling the development of ecosystem models that directly relate to both ecosystem condition and the ecological functions that protect humans, such as from potentially degraded water quality and flooding events, at a regional scale. The approaches presented in the two case studies in Chapter 5 follow the same general principles of landscape ecology put forth in Chapters 2 and 3, and build upon the lessons learned in the case studies of this chapter, through the development and application of improved modeling techniques.

5

Applied Analyses of Regional Scale Landscape Gradients of Condition

Two final case studies are discussed in this chapter, each of which build upon the lessons learned from the previous work discussed in Chapter 4. In this chapter, we build further on the basics of metric development and application across broad landscapes (i.e., the Great Lakes project), and the use of landscape models to analyze the specific impacts of environmental change on those broad-scale landscape elements (i.e., the California and North Carolina Coastal project), and focus on measurable outcomes that are of keen importance to decision makers, such as water quality and water quantity, at a landscape scale. One of the following studies elucidates water quality metrics and indicators in the Ozark Mountains and the other case study elucidates water quantity metrics and indicators in Missouri River and Mississippi River Watersheds, given specific changes across the landscape, with an added special focus on the Kansas River, one of the 22 tributaries in the Missouri/Mississippi River case study. The pioneering work in this chapter utilizes remote sensing and field-based data to develop GIS models to determine the drivers of change, similar to those projects in the previous chapter, yet draws out more of the details of the processes involved in each of the models, specifically in terms of water chemistry, aquatic microbiology, and hydrology. The projects in this chapter also expand on the previous chapter's discussion and work by providing additional details on the specific actions that decision makers can make in the geographies analyzed, and elsewhere worldwide where conditions may be similar, to better serve the needs of communities (Lopez in preparation). You will find that watershed analyses, such as those described in this chapter, are an increasingly important priority for local and regional planners, and the projects in this chapter will assist you, the reader, in several invaluable examples of approaches to utilize in other project areas that may share similar needs.

5.1 Case Study: Ozark Mountains Watershed

We continue to build on the lessons learned from previous chapters by providing a detailed example in the Ozark Mountains of the United States, where remote sensing imagery and landscape ecology have come together to model watershed gradients among multiple areas of the landscape, providing the baseline for statistical inference of conditions across the landscape with regard to water chemistry and aquatic microbiology.

5.1.1 Ozark Mountains: Landscape Context and Background

The Ozark Mountains, located in Arkansas and Missouri, are a relatively unique landscape that contain the headwaters of the White River, and a multitude of smaller tributary streams that flow to the White River's main stem, which then continues downstream from the mountains into the highly agricultural Mississippi Alluvial Plain, and ultimately to its confluence with the Mississippi River, and on to the Gulf of Mexico.

A fairly important characteristic of the Ozarks to recognize, and a challenging one for the use of remote sensing, is the fact that a large proportion of water flow in the region is underground, given the Ozark's karst topography across much of the study area. *Karst* refers to a distinctive topography that exists across approximately 10% of the Earth's surface, and 25% of water relied upon by humans worldwide. The formation karst results from the dissolution of underlying soluble rocks by surface water or groundwater. Although karst topography is a relatively unique landform in the continental United States, it is also present in many other landscapes around the world, from Hawai'i to Laos, South Africa to Albania, and in other areas all around the globe. Understanding the conditions of karst in the Ozarks and elsewhere is important because of the relevance to determining the vulnerability of water resources to water pollution.

The Ozarks project is in a 21,848 km² area of land that encompasses the headwaters of the White River and generally the Ozark Mountains (Figure 5.1). The study area contains a mix of agriculture (principally poultry production with additional contributions from cattle and hay operations), forested, urban land cover, and aquatic areas. The White River originates in northwestern Arkansas and flows through southwestern Missouri and north-central Arkansas. The White River then descends from the Ozark Mountains into Arkansas' agricultural plain, where it meanders to its confluence with the Mississippi River. There are seven major dams that maintain reservoirs on the White River, which exert major biophysical influences on the river and associated ecosystems of several sensitive natural areas, such as National Wildlife Refuges, National Forests, and a National Scenic River.

Although primarily a rural area, during the past half-century the Ozarks have undergone relatively large population increases, along with land cover and associated environmental changes that accompany human activities (Lopez et al. 2006b). Changes in Ozarks land cover often accompany growth, including urbanization (Figure 5.2a) and the clearing of forestland (Figure 5.2b), which may result in biogeochemical effects on streams (Bormann et al. 1968; Meyer and Likens 1979; Vitousek and Melillo 1979; Peterjohn and Correll 1984).

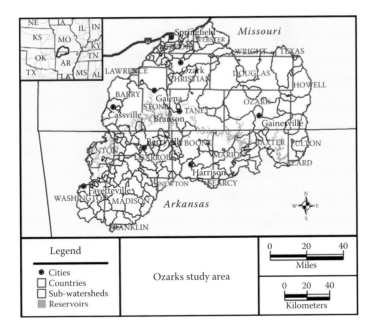

FIGURE 5.1
The Ozark Mountains study area (21,848 km²), covering multiple counties in Arkansas and Missouri. Boundaries for the 244 customized sub-watersheds used for analyses in this study are shown for reference to their location, relative to study area counties and cities. The main stem of the White River (including reservoirs) is shown in blue.

Such changes in the Ozarks have raised concerns by local communities, watershed management groups, and regulators at the USEPA about surface water quality, specifically eutrophication and bacterial contamination in the White River Watershed. Because of potentially deleterious effects on human health, recreational values, and ecological functions of the aquatic ecosystems of this unique region of the country, watershed assessment efforts are an ever-increasing priority for water quality resource managers and regulators in the region (Messer et al. 1991; Haggard et al. 2003; Turner and Rabalais 2004; Radwell and Kwak 2005). Broad-scale landscape analyses in other areas of the United States suggest that increases in the presence and intensity of human land use or the loss of natural land cover may result in relatively increased nutrient and bacteriological loading to surface water (May et al. 1997; Jones et al. 2001, 2004; Lopez et al. 2003; Mehaffey et al. 2005; Riitters and Coulston 2005).

Despite the apparent concerns about, and interest in, the water quality of the Ozarks, there are still several unmet needs for a thorough understanding of how the presence of agricultural land, urban areas, and forested land in the riparian areas and the watershed as a whole may affect surface water conditions. Some of the answers to the questions about landscape conditions' effect upon water quality require a broad-scale approach. Notwithstanding

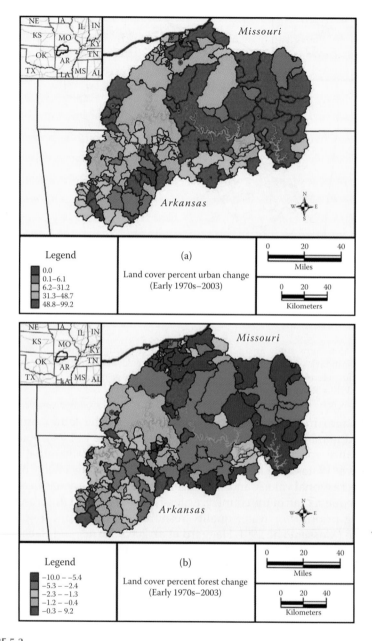

FIGURE 5.2
Land cover change (early 1970s to 2003) among 244 customized sub-watersheds for the Ozark study area. (a) Percent urban change. (b) Percent forest change. The main stem of the White River (including reservoirs) is shown in blue.

the many biophysical and social challenges of the region, remote sensing techniques were brought to bear in the Ozarks to address these important questions, providing an invaluable tool, to address these critical landscape ecology issues of the Ozarks, specifically with regard to water pollution. To accomplish the goals of the remote sensing approach, landscape metrics (e.g., agricultural, urban, and forestland cover metrics) were analyzed and tested for their correlation with water quality parameters in associated surface waters at the sub-watershed scale at three finer scales (i.e., within three progressively narrower riparian zones: 120 m, 30 m, and directly adjacent to stream and river banks).

To assess watershed condition, 46 of the broad-scale landscape detailed metrics commonly utilized in the late 1980s and 1990s (Forman and Godron 1986; Forman 1995; Gustafson 1998; O'Neill et al. 1999; Riitters et al. 2000; Jones et al. 2001) were used, which all fall within the general types outlined in Chapter 2; the 46 metrics were used to predict total phosphorus (TP) concentration, total ammonia (TA) concentration, and *Escherichia coli* (*E. coli*) cell counts at a sub-watershed's outfall location, among 244 sub-watersheds of the White River, which define the Ozarks from a human water-use standpoint, as well as an aquatic ecosystem perspective. The three water quality parameters selected are important from a human health perspective, and were the dependent variables (response parameters) commonly used to measure surface water conditions throughout the United States.

Among the human health and well-being water quality parameters assessed using the remote sensing approach were total phosphorus (TP) concentration, total ammonia (TA) concentration, and *E. coli*. TP was selected as a surface water response parameter because freshwater ecosystem productivity is nutrient limited, and these ecosystems may become eutrophic when an excess input of phosphorus (P) and nitrogen (N) is converted to biomass by aquatic organisms. Eutrophication symptoms have been increasingly observed in certain lakes and streams of the Ozark Mountains, including poor water clarity, odors, poor taste, low oxygen content, a change in the presence and abundance of fish, the presence of dead fish, increased algal growth, and the presence of harmful cyanobacteria (Lory 1999). The Ozarks are actually well known for their oligotrophic lakes and streams, which is generally a result of the relatively older soil structure and relatively few crop activities high in the watershed as a whole. There are, however, possible sources of P in the Ozark Mountains, including industrial effluent and wastewater treatment outfall and nonpoint source agricultural field runoff, urban runoff, leaks from septic systems associated with human habitation, and activities at locations that contain cattle, poultry, or swine wastes (USEPA 1997; USGS 1999). Poultry production is particularly prevalent in the study area within Northwest Arkansas and Southwest Missouri. For nitrogen, ammonia (NH_3 and NH^+) was specifically selected as a surface–water response parameter because it is present in variable concentrations in many surface and groundwater supplies.

A product of microbiological activity, ammonia, when found in surface water, is regarded as indicative of sanitary pollution. In the presence of dissolved oxygen, ammonia is rapidly oxidized by specific bacteria to nitrite and nitrate. Ammonia, being a source of N, is also a nutrient for algae and other forms of plant life and thus contributes to the overloading of natural systems and is a cause of pollution. Ammonia can be toxic to aquatic life and at specific concentrations has been found to be a source of toxic effects to aquatic life in some streams. Inadequately treated municipal wastewater, agricultural runoff, and groundwater contamination by fertilizer, storm water, and feedlots are potential sources of ammonia (and nitrate) to streams. The major potential sources of ammonia in the Ozarks are inadequately treated wastewater, agricultural runoff, water that contains fertilizer, storm water, and agricultural feedlots (USEPA 1995).

Escherichia coli is a species of fecal coliform bacteria that is specific to fecal material from humans and other mammals and birds. *E. coli* was selected as a surface water response parameter because the EPA considers it to be one of the important indicators of health risk from water contact in recreational waters (USEPA 1997). *E. coli,* as well as other fecal coliform and *streptococci* bacteria, have long been indicators of sewage contamination in streams. *E. coli* also indicates the presence of other pathogenic bacteria, viruses, and protozoans that live in the digestive systems of humans and other animals. Symptoms of elevated levels of *E. coli* in surface water are cloudy water, unpleasant odors, and increased oxygen demand. Sources of *E. coli* contamination in surface water include municipal wastewater treatment plants, ineffective septic systems, domestic animal manure, wild animal feces, and storm water runoff (Lory 1999).

Total phosphorus, TA, and *E. coli* measurements of surface water from field sampling programs over the past decades were necessary but were costly and labor-intensive, and yet a full landscape view of predicted water quality is unattainable with even this massive field effort. Consequently, as is the case in many watersheds, the impressive total number of samples was still relatively small in comparison to what is necessary to accomplish a synoptic assessment of the larger landscape, as a whole. As is also the case in other watersheds, data sets were collected over a period of years (in this analysis, 6 years) and contained missing values, likely the result of cost limitations and differing goals of the projects that were responsible for collecting the data. These pitfalls of data availability, as was an impediment outlined earlier in Chapter 4, are the subject of the next section, where techniques for overcoming this pitfall are discussed and applied.

5.1.2 Determining Watershed Conditions

As with many other regional ecological data sets, field-based data in the Ozarks are only partially valid for statistical analyses due to a variety of missing or other problems that stem from their variability in collection approach, timing, and spatial distribution. Because standard regression analyses are sensitive to missing values and (auto)correlation of predictors, they create a problem for their appropriate use in landscape metric statistical analyses. To overcome these pitfalls, partial least squares (PLS) analysis was intentionally applied across the study area, particularly because this technique is useful for analyzing data sets with missing values, collinear variables, and small sample sizes (Wold 1995; Nash et al. 2005), all of which apply to the available data sets in the Ozarks, and a result of attempting to analyze such a large extent of the landscape. On the positive side, such a broad landscape approach and the availability of remote sensing data, and thus the ability to generate landscape metrics, allowed for the overall accomplishment of broader landscape project goals. Specifically, low prediction error and high prediction accuracy of PLS was used to predict (i.e., to develop) landscape indicators of nutrient concentrations and bacterial counts in surface water of the Ozark Mountains.

Initially, surface water measurements from 1997 to 2002 were compiled from U.S. Geological Survey and state agency data sets, resulting in 244 stream sample locations (Figure 5.3a). A major assumption of these watershed analyses was that, because of the Ozark's karst topography, subsurface flow of water may be substantial and relatively constant across the study area and thus would not affect the relative surface water quality results among the 244 sub-watersheds.

The 244 surface water sampling locations (Figure 5.3a) were used to delineate 244 sub-watersheds (Figure 5.3b) of the White River, such that the total area contributing to overland flow to the sample point was delineated (Van Remortel et al. 2004, 2005). Thus, each of the 244 sample points was positioned as the single outlet (hereafter also referred to as a pour point) from a given sub-watershed. The 244 sub-watersheds provided quantifiable contributing areas of runoff where landscape metrics could be calculated. Some of the 244 sub-watersheds were nested within other larger sub-watersheds, and thus the total area of the 244 sub-watersheds exceeded the Ozark study area (i.e., 21,848 km^2). The Missouri Resource Assessment Partnership (2008) and the Arkansas GAP Analysis Project (2008) land cover data were evaluated using ArcGIS geographic information systems software (v.9; Environmental Systems Research Institute, Inc., Redlands, California). While evaluating the state land cover data, it was determined that the original classification did not accurately depict 2003 conditions and that there were inconsistencies between states. Thus, 2003 land cover was updated by a combination of supervised classification of contemporaneous Landsat Enhanced Thematic Mapper (ETM+) data (Figure 5.3c) and standard photo-interpretation techniques with 1-m spatial resolution (2003) Digital Ortho Quarter Quadrangle photography (Lillesand et al. 2004) using Imagine remote sensing image processing software (v.9.0; Leica Geosystems Geospatial Imaging, LLC, Atlanta, Georgia).

Classification and photo interpretation resulted in a customized 2003 land cover data set (Figure 5.3d) that was used to calculate 46

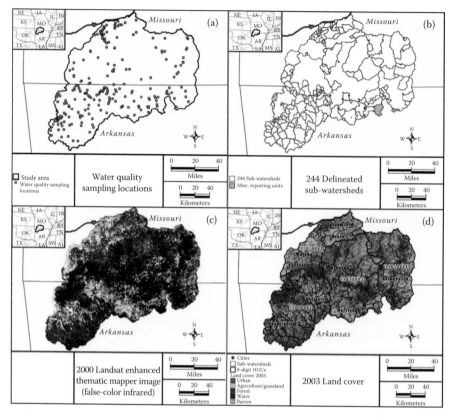

FIGURE 5.3
Ozarks Mountains (Missouri and Arkansas) study area, where 244 water quality sampling locations were sampled and used as analysis "pour points" (a), from which 244 contributing sub-watersheds were delineated (b). Circa 2000, multiple Landsat Thematic Mapper satellite data sets (c) and 2003 digital aerial photography was used to produce a current (summer 2003) land cover map of the study area (d). The 2003 land cover was used to calculate 46 broad-scale landscape metrics.

commonly used landscape metrics (Table 5.1). The five land cover classes are as follows:

1. Forest (e.g., nature reserves and woodlots)
2. Urban (e.g., urban centers and small rural towns)
3. Water (e.g., streams, rivers, lakes, and ponds)
4. Agriculture/grassland (e.g., pasture and animal feedlots, poultry operations, row cropfields, golf courses, or other similar areas of herbaceous vegetation)
5. Barren (e.g., gravel pits and sandy beaches)

TABLE 5.1

Forty-Six Landscape Metrics Used in the Ozark Study Area

Landscape Metric Description	Abbreviation	Applicable Landscape Unit
Sub-watershed total area	delin_Area	Sub-watershed
Sub-watershed total perimeter	delin_Perimeter	Sub-watershed
Minimum topographic elevation	Elevmin	Sub-watershed
Maximum topographic elevation	Elevmax	Sub-watershed
Mean topographic elevation	Elevmean	Sub-watershed
Variation in topographic elevation	Elevstd	Sub-watershed
Total range of topographic elevation	Elevrange	Sub-watershed
Total stream length	Strmlen	Sub-watershed
Total stream density	Strmdens	Sub-watershed
Forest patch count	Fnumber	Sub-watershed
Forest patch density	Fdensity	Sub-watershed
Area of largest forest patch	Flargest	Sub-watershed
Percent of entire sub-watershed comprised by largest patch of forest	F_plgp	Sub-watershed
Mean area of forest patch	Favgsize	Sub-watershed
Mean distance of the closest patch of forest	F_mdcp	Sub-watershed
Number of forest patches with neighbors—210-meter search radius	F_pwn	Sub-watershed
Forest edge—edge width defined as 210 meters	Fedge210	Sub-watershed
Forest core—edge width defined as 210 meters	Fcore210	Sub-watershed
Edge-to-area ratio—edge width defined as 210 meters	F_e2a210	Sub-watershed
Percent riparian forestland cover—adjacent to streams and rivers	Rfor0	Sub-watershed
Percent riparian forestland cover—within 30 meters of streams and rivers	Rfor30	Riparian zone of sub-watershed
Percent riparian forestland cover—within 120 meters of streams and rivers	Rfor120	Riparian zone of sub-watershed
Percent forestland cover	Pfor	Sub-watershed
Percent barren land cover—adjacent to streams and rivers	Rmbar0	Sub-watershed
Percent barren land cover—within 30 meters of streams and rivers	Rmbar30	Riparian zone of sub-watershed

(Continued)

TABLE 5.1 (CONTINUED)

Forty-Six Landscape Metrics Used in the Ozark Study Area

Landscape Metric Description	Abbreviation	Applicable Landscape Unit
Percent barren land cover—within 120 meters of streams and rivers	Rmbar120	Riparian zone of sub-watershed
Simple diversity index (i.e., richness) of land cover types	S	Sub-watershed
Shannon–Wiener Index of land cover types	C	Sub-watershed
Simpson's Diversity Index of land cover types	H	Sub-watershed
Percent riparian agricultural (including all cropland and pasture/grassland) land cover—adjacent to streams and rivers	Ragt0	Riparian zone of sub-watershed
Percent riparian agricultural (including all cropland and pasture/grassland) land cover—within 30 meters of streams and rivers	Ragt30	Riparian zone of sub-watershed
Percent riparian agricultural (including all cropland and pasture/grassland) land cover—within 120 meters of streams and rivers	Ragt120	Riparian zone of sub-watershed
Percent agriculture (including all cropland and pastureland/grassland) land cover	Pagt	Sub-watershed
Percent agriculture (including all cropland and pastureland/grassland) land cover on slopes greater than 3%	Agtsl3	Sub-watershed
Percent barren land cover	Pmbar	Sub-watershed
Percent riparian urban land cover—adjacent to streams and rivers	Purb0	Riparian zone of sub-watershed
Percent riparian urban land cover—within 30 meters of streams and rivers	Purb30	Riparian zone of sub-watershed
Percent riparian urban land cover—within 120 meters of streams and rivers	Purb120	Riparian zone of sub-watershed
Percent urban land cover	Purb	Sub-watershed
Percent impervious surfaces based upon roads	Pctia_rd	Sub-watershed
Total road length	Rdlen	Sub-watershed
Total road density	Rddens	Sub-watershed

Source: USEPA, ATtILA User Guide, EPA/600/R-04/083. USEPA Washington, DC, 2004b.

The updated Ozark land cover map was exported as an (ESRI) ArcGIS (v. 9.0) grid. The Ozark land cover grid was also independently field verified (i.e., blind field visits to classified grid-cell locations) during the summer of 2005. The land cover classification maps resulted in a 96% user's accuracy (n = 73).

5.1.3 Determining Patch Characteristics with Geospatial Statistics

Stream water samples among locations in watersheds were (previously) collected and archived in the U.S. Environmental Protection Agency's STORET/ Legacy Data Center (USEPA 2008) or the USGS National Water Inventory System (USGS 2008). Archived data were previously analyzed for TP, TA, and *E. coli* in the laboratory from samples collected in the field at the 244 sub-watershed pour points (Figure 5.3a) from 1997 through 2002. Forty-six broad-scale landscape metrics (Table 5.1) were calculated for each sub-watershed (Figure 5.3b). Mean values for TP concentration (mg L^{-1}), TA concentration (mg L^{-1}), and *E. coli* count (bacterial cells/100 mL) were used in a PLS analysis. The analyses were conducted on data from a limited number of sites (n = 18, 15, and 6 sub-watersheds, respectively) to predict sub-watershed values for each water quality variable among the 244 sub-watersheds of the study area. The limited number of sub-watersheds used in the PLS analyses was the result of data limitations in the water sample data archive. Although many of the 244 sub-watersheds contain some of the smaller sub-watersheds, none of these "overlapping sub-watersheds" were used in the PLS analyses.

Principal component analyses use one data set to compose few orthogonal (latent) variables, summarizing information from many variables into a data set. Partial least squares analyses, in contrast, use two data sets to summarize information and provide subsequent predictions. The PLS analysis method is based on first computing relevant projections (i.e., linear combinations) of the predictors X and then using these new variables in a regression equation for predicting the response(s) Y. The result is a matrix of components (or scores, which are often denoted by T) that represents the most important information in X, for predicting Y, and is thus used in the regression equation instead of the original X variables themselves. It can be shown that the scores are orthogonal.

Partial least squares analyses produced n − 1 factors, where the first factor accounts for most of the covariance between response and landscape metrics. The second factor was determined by extraction of the residuals from the first and resolved the linear combinations of both data sets such that their covariance was maximized. This process was repeated by taking residuals from each previous factor, producing n − 1 factors, where n was the number of observations. For example, if the number of sample sites (observations) used in the analysis was 18, then 17 factors were used.

After defining the significant PLS factors, and for interpretation, the scores and weights were computed and plotted in simple scatter plots. The strength of relationship patterns between Y and X scores, the influence, and any clustering or irregular pattern in sites can be seen by plotting of Y and X scores. The weights are the regression coefficients of the variables in X and Y regressed on the various variables in T, which represent the contribution of each variable. The variable influence on projection (VIP) was also used as a measure of the relative contribution of each of the landscape metrics on the response variable. The weight of each landscape variable and percent variability in response are incorporated into the VIP calculation. Regression coefficient values indicate the contribution of each predictor for response. The VIP value is based on response and predictor measures. Therefore, if the VIP for a predictor is below a threshold value, it implies that variable has a relatively small contribution to the prediction and may be deleted from the PLS model. Variables with a VIP value less than a 0.8 threshold are customarily considered small contributors (Wold 1995). An improved model can be built by including variables with high VIP values and excluding others with a low VIP.

All of these methods were applied as an approach for overcoming the general paucity of field-based data available to validate the landscape ecological models, allowing for better calibration of models that are derived from the land cover data utilized in this project. Accordingly, 18 non-nested watersheds for the TP PLS model were used, 15 non-nested watersheds for the TA PLS model, and 6 non-nested watersheds for the *E. coli* PLS model. The minimum and maximum values measured for each of the three field-measured surface water parameters (Table 5.2) were within the range reported in this and other similar regions of the United States.

The TP model resulted in one significant factor explaining 91% of the variability in TP concentration at a sub-watershed's pour point (Table 5.3). Several key landscape variables in the TP model are positively correlated

TABLE 5.2

Total Phosphorus (TP), Total Ammonia (TA), and *E. coli* Partial Least Squares (PLS) Model Prediction Results and Actual Field Measurements within the Ozark Mountains Watershed

	PLS Predicted Values (N = 244)		Actual Measurements (n = 18, 15, 6 for TP, TA, *E. coli*, respectively)	
	Minimum	Maximum	Minimum	Maximum
TP	0.0344	0.4186	0.022	2.181
TA	0.0035	0.1498	0.012	0.122
E. coli	0.6842	20397	24.58	20620
Group	**TP**	**TA**	**E. coli**	
1	0.004 – ≤ 0.052	0.034 – ≤0.073	0 – ≤2000	
2	>0.052 – ≤0.101	>0.073 – ≤0.150	2000 – ≤6000	
3	>0.101 – ≤0.150	>0.15	>6000	

TABLE 5.3

Coefficients of the Non-Centered (i.e., Model Output) Value of Landscape Metrics That Are Predictive of the ln (Total Phosphorous [TP]), Total Ammonia (TA), and ln (*E. coli*) within the Ozark Mountains Watershed

Landscape Metrics	TP		TA		*E. coli*	
	Coefficient	VIP	Coefficient	VIP	Coefficient	VIP
Intercept	−1.514		0.077		−2.255	
Fedge210	−0.10327	0.978				
F_mdcp	0.00100	0.843				
F_plgp					−0.00783	1.010
Rfor0	**−0.12393**	**1.028**	**−0.00020**	**0.914**	**−0.00026**	**0.981**
Rfor30	**−0.11484**	**1.015**	**−0.00019**	**0.923**	**−0.00066**	**0.983**
Rfor120	**−0.10367**	**1.031**	**−0.00017**	**0.935**	**0.00175**	**0.980**
Pfor	**−0.08166**	**0.969**	**−0.00018**	**0.971**	**0.00741**	**0.9795**
Rurb0	**0.000036**	**1.037**	**0.00034**	**1.065**	**0.00907**	**1.012**
Rurb30	**0.000033**	**1.039**	**0.00031**	**1.064**	**0.00851**	**1.013**
Rurb120	**0.000027**	**1.056**	**0.00027**	**1.063**	**0.00722**	**1.006**
Purb	**0.000017**	**0.985**	**0.00023**	**1.061**	**0.00564**	**1.010**
Rhum0	0.000020	1.036			0.00026	0.981
Rhum30	0.000018	1.014			0.00066	0.983
Rhum120	0.000016	1.017			−0.00175	0.980
Pctia_rd	0.000062	0.945			0.01363	1.019
Rddens	0.001521	0.934			0.10008	1.043
Pmbar	0.168738	1.071				
Rmbar120	0.169849	1.064			0.00195	0.763
Strmdens	−0.128696	0.946			−0.90015	1.016
Elevmin					0.00024	1.215
Number of factors	1		1		2	
% Variation	91		93		99.7	

Note: Number of significant PLS factors and percent variation explained by PLS for the responses are in the last two rows. Bolded values are for metrics that were commonly used among TP, TA, and *E. coli*.

with TP concentration (mg L^{-1}): the sub-watershed mean minimum (inter-forest) distance to the closest patch of forest, the proportion of urban land adjacent to streams and in the 30- and 120-m riparian zones, the proportion of urban land within the sub-watershed, the proportion of human land use adjacent to streams and in the 30- and 120-m riparian zones, the proportion of impervious surfaces and road density within the sub-watershed, and the proportion of barren land in the 120-m riparian zone and within the sub-watershed. Several key landscape variables in the TP model are negatively correlated with TP concentration (mg L^{-1}): the amount of forest edge in the sub-watershed, the proportion of riparian forest adjacent to streams and in the 30- and 120-m riparian zone, the proportion of forestland within the sub-watershed, and sub-watershed stream density.

The TA model resulted in one significant factor explaining 93% of the variability in surface water TA concentrations (Table 5.3). Several key landscape variables in the TA model are positively correlated with TA concentration (mg L^{-1}): the proportion of urban land adjacent to streams and in the 30- and 120-m riparian zones and the proportion of urban land within the sub-watershed. Several key landscape variables in the TA model are negatively correlated with TA concentration (mg L^{-1}): the proportion of riparian forest adjacent to streams and in the 30- and 120-m riparian zone and the proportion of forestland within the sub-watershed.

The *E. coli* model resulted in two significant factors explaining 99.7% of the variability in *E. coli* cell counts at a sub-watershed's pour point (Table 5.3). Several key landscape variables in the *E. coli* model are positively correlated with *E. coli* cell count (bacterial cells/100 mL): the proportion of riparian forest in the 120-m riparian zone, the proportion of forestland within the sub-watershed, the proportion of urban land adjacent to streams and in the 30- and 120-m riparian zones, the proportion of urban land within the sub-watershed, the proportion of human land use adjacent to streams and in the 30-m riparian zone, the proportion of impervious surfaces and road density within the sub-watershed, the proportion of barren land in the 120-m riparian zone, and the sub-watershed elevation minimum.

Several key landscape variables in the *E. coli* model are negatively correlated with predicted *E. coli* cell count (bacterial cells/100 mL): the proportion of largest forest patch to total watershed forest area, the proportion of riparian forest adjacent to streams and in the 30-m riparian zone, the proportion of human land use in the 120-m riparian zone, and sub-watershed stream density.

Total P and TA models have similar spatial trends for percent urban and percent forest (eight metrics in italics in Table 5.3). For each land cover type, the coefficient adjacent to streams (i.e., Rurb0 or Rfor0) is greatest, with decreasing correlations for wider riparian zones (e.g., progressively from Rfor0, to Rfor30, to Rfor120, to Pfor). The spatial trends are similar for the *E. coli* models, except the negative correlations for Rfor120 and Pfor modeled for TP and TA, which are positive correlations for *E. coli*. Positive PLS correlations that are shared between the TP model and the *E. coli* model are those for the percentage of human use in riparian zones, road density, and the percentage of impervious surfaces resulting from roads within a sub-watershed. The percent of barren land in riparian zones is also an important factor for the TP and the *E. coli* models but is less influential on the *E. coli* model than on the TP model (Table 5.3). The minimum elevation in a sub-watershed is also an important positive correlate for the *E. coli* model, but the correlation is very weak. Negative correlates in the *E. coli* model include the percentage of a sub-watershed that the largest patch of forest represents, stream density in a sub-watershed, and percentage of human land use within 120-m of streams.

The PLS model results predict the highest TP concentration in sub-watersheds in the vicinity of Springfield and Cassville, Missouri, and Harrison, Arkansas; to the east of Berryville, Arkansas; and in western Fulton County, Arkansas (Figure 5.4a). The PLS model results predict highest TP concentrations at the pour point of sub-watersheds in the vicinity of Springfield and Cassville, Missouri, and Harrison, Arkansas; to the east of Berryville, Arkansas; western Fulton County, Arkansas; central Baxter County, Arkansas; and a 9-km^2 sub-watershed in central Madison County (vicinity of Huntsville), Arkansas (Figure 5.4a). The PLS model results predict highest TA concentrations at the pour point of sub-watersheds in the northwestern portion of the study area; in the vicinity of Harrison, Arkansas; to the east of Berryville, Arkansas; western Fulton County, Arkansas; central Baxter County, Arkansas; the same 9-km^2 sub-watershed in central Madison County (vicinity of Huntsville), Arkansas, as found for TP; two other sub-watersheds crossing the Madison–Benton and Madison–Washington county lines, Arkansas; southern Howell County, Missouri; and two small sub-watersheds in the northeastern portion of the study area (Figure 5.4b). The PLS model results predict highest *E. coli* cell counts concentrations at the pour point of sub-watersheds in the vicinity of Springfield and Cassville, Missouri; in the far southern portion of the study area; and in the eastern Webster county, Webster–Douglas county, and Webster–Douglas–Christian county, Missouri, regions (Figure 5.4c).

Because the 244 sub-watersheds in the study area are overlapping and often nested within each other, the geographic distribution of PLS model predictions for TP, TA, and *E. coli* are difficult to fully display in a two-dimensional map form (Figure 5.4c). For this reason, a personal computer visualization tool was developed and released to the public to accompany the mapped PLS model results, allowing watershed managers and other users to best disseminate and explain the multiscale results (Figure 5.5; Lopez et al. 2006b).

Total P, TA, and *E. coli* PLS predicted values were used to determine the cumulative predicted vulnerability of sub-watersheds in the study area. The combined vulnerability of ecosystems was graphed using three axes (TP concentration, TA concentration, and *E. coli* cell count [Figure 5.6]) and qualitatively grouped into four (predicted) parameter states:

1. Sub-watersheds that have high TP concentrations, high TA concentrations, and high *E. coli* cell counts
2. Sub-watersheds that have low TP concentrations, high TA concentrations, and high *E. coli* cell counts
3. Sub-watersheds that have moderate TP and TA concentrations and moderate *E. coli* cell counts
4. Sub-watersheds that have low TP and TA concentrations and moderate *E. coli* cell counts

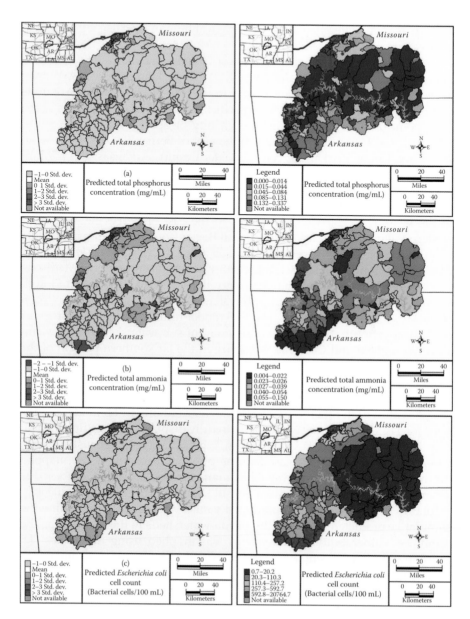

FIGURE 5.4
Two-dimensional depiction of the geographic distribution of partial least squares model predictions for (a) total phosphorus, (b) total ammonia, and (c) *E. coli* provide a general view of the relative vulnerability of watersheds but because watersheds are inherently nested, some areas of the 244 sub-watersheds are obscured in this figure. Smaller nested sub-watersheds are layered on top of the larger sub-watersheds such that a portion of all 244 sub-watersheds are depicted in this two-dimensional map. The main stem of the White River (including reservoirs) is shown in blue.

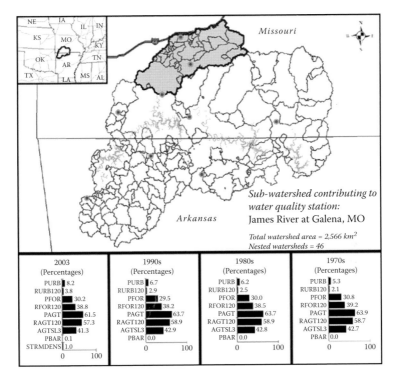

FIGURE 5.5
A personal computer visualization tool has been provided to users of the total phosphorus, total ammonia, and *E. coli* partial least squares model results for the Ozarks that allows for the interpretation of nested, overlapping sub-watersheds. The example shows the use of the visualization tool to explore the landscape metrics in 1 of the 244 customized sub-watersheds (the 2,566 km² James River at Galena, Missouri, sub-watershed), which contains 46 other smaller "nested sub-watersheds" that can be similarly viewed using the visualization tool.

5.1.4 Inferring Ecosystem Water Quality Conditions Using Riparian Landscape Metrics

The trends for the TP, TA, and *E. coli* models are the same for four specific urban landscape metrics (Rurb0, Rurb30, Rurb120, and Purb), indicating similar sub-watershed scale water quality responses. The trends observed among the three PLS models for TP, TA, and *E. coli* also indicate sub-watershed scale water quality responses to the four specific forest landscape metrics (Rfor0, Rfor30, Rfor120, and Pfor) but with trends and magnitudes that are different for *E. coli* than for the other two water quality parameters. The TP and *E. coli* models were more complex than the TA model in that the former two models include other statistically and ecologically significant landscape variables.

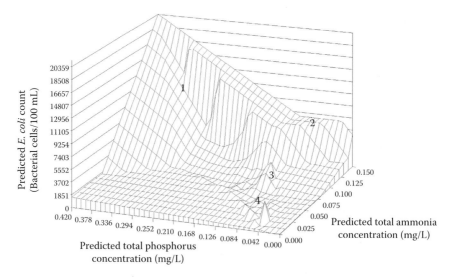

FIGURE 5.6

Three-dimensional plot of predicted total phosphorus concentrations, total ammonia concentrations, and *E. coli* cell counts among 244 sub-watersheds in the Upper White River region of the Ozarks. Graphic integration of the predicted values generally depicts (1) most vulnerable sub-watersheds, (2) highly vulnerable sub-watersheds, (3) moderately vulnerable sub-watersheds, and (4) least vulnerable sub-watersheds.

5.1.4.1 Total Phosphorus Model

Key contributors to the TP PLS model are percent barren (positive correlation) and stream density (negative correlation). Greater stream density increases the likelihood of TP interception by surface water but also provides a greater volume of water to dilute the concentration effects of runoff. Increased stream density may also increase the likelihood of biological interception and thus uptake, accumulation, and transformation of TP during biological and related biophysical processes. The forest-related variables generally indicate that increased percentage of forest (i.e., Rfor0, Rfor30, Rfor120, and Pfor) reduces the amount of P in streams and that increased fragmentation of forests (i.e., Fedge210 and F_mdcp) increase the amount of TP in streams. Conversely, the barren-related variables indicate that a decrease in vegetated cover increases the amount of P in streams (Table 5.3). These results are in keeping with the general observations that P is transported in runoff, often from agricultural fields into streams, and that it can travel overland attached to particles of soil or manure into streams, and this process is facilitated when fewer physical impediments (e.g., trees, shrubs, and other vegetation) to runoff flow exist.

Because fine soil particles have a large capacity for immobilizing phosphorus in the substrate, TP in soil is even higher in concentration than the soluble phosphorus content. The capacity of sandy soils to adsorb phosphorus is limited and water between sand particles can be more easily saturated with TP than in organic or clayey soils. At a landscape scale, it was found that barren areas (either sandy beaches on lakes or rivers, urban- or road-adjacent cleared areas, or surface mining areas) were positively correlated (Table 5.3) with TP at the sub-watershed scale (Pmbar) and within 120 meters of streams (Rmbar120). The indirect effects of agriculture (i.e., manure in pastures, the use of animal manure in cropland, and manufactured fertilizer use in cropland) on TP in streams was not detected using landscape metrics, perhaps because of the relatively low intensity of inputs from cropland in the Ozarks, as compared to other dominantly cropped areas like the Mississippi Alluvial Valley (lower White River Watershed). Research from Arkansas on poultry litter and swine manure applied to pastures shows that soluble phosphorus concentrations increase in direct proportion to increasing application rate and that soluble phosphorus dominates the content of "flash phosphorus loss events" (Lory 1999), which occur rapidly, perhaps too fast to measure using averaged field data over the study period.

Creating (or restoring) vegetated areas between agricultural fields and water resources (sometimes called "buffers") may lower P concentrations in streams, as suggested by our landscape metric PLS models. Specifically, our results suggest that those areas that have non-vegetated areas along streams are predominantly urban land cover and that those areas that have forest vegetation within 120-m of streams are responsible for lower concentrations of TP in the associated sub-watersheds. Human land use in the TP model (positive correlation) is not as strong as the urban metrics, which is likely an artifact of the necessity of combining natural grassland with agricultural land in our land cover classes. Also, natural grassland (i.e., grassland not used for grazing) was presumed to be a very small proportion of the landscape in the Ozarks; thus, human land use in the riparian zone of 120-m (i.e., Rhum0, Rhum30, Rhum120) is mainly the additive combination of all agricultural land (which was not correlated with TP by itself) and all urban land. An important finding is that the closer the urban land cover or (combined) human land use is to the stream bank within a given sub-watershed (i.e., non-forested areas) the greater the effect on TP concentrations, as observed at the pour point of the sub-watershed. It may also be that the effects of agricultural areas (as a sub-component of the human land use metrics) on TP concentrations are dramatically ameliorated when agricultural land is a distance greater than 120-m, perhaps because of the greater opportunities for soil adsorption and biological accumulation between the TP sources and the streams of the sub-watershed. Forest percentage metrics in the riparian areas and the sub-watershed consistently indicate that loss of forest results in an increase in TP and that these correlations are stronger within closer proximities to the stream. For example, for forest metrics, the lowest TP model coefficient is for percent forest within the sub-watershed, and the greatest TP model coefficient is for forest adjacent to stream (Table 5.3).

The TP model results for forest edge (Fedge210) are consistent with the other forest metric results, suggesting that forest patches that have greater edge lengths (i.e., irregularly shaped and undulated patches vs. uniformly shaped patches) throughout a sub-watershed decrease the concentrations of TP at the pour point of a sub-watershed. Greater forest edge and patch size metrics may indicate increased runoff interception opportunities for TP; interception of soil particles and other particles onto which TP may adsorb reduces the rate and total amounts of TP arriving at streams. Also, riparian forested areas tend to have greater edge lengths than forest patches that are interior to the sub-watershed, indicating greater riparian forest in sub-watersheds with greater forest edge length, and thus a greater reduction in TP in those sub-watersheds with greater percentages of riparian forest. Similar to edge length of forest patches, this may be the result of the decreased opportunity for TP in runoff to be intercepted by forest vegetation and soils in landscapes where forest patches are far from each other, tending to increase the likelihood that TP in runoff would eventually flow into streams of the more forest-fragmented sub-watersheds. The results suggest that the loss of forested areas at the riparian and sub-watershed scales, along with replacement of these forested areas with urban land cover, agricultural land cover (including the associated activities of inadequately treated municipal wastewater, agricultural runoff, feedlots, and storm water runoff), or other non-vegetated land cover types are dominant sources of TP to streams in the Ozarks.

5.1.4.2 Total Ammonia Model

The results of the TA PLS model are the clearest in terms of the TP model discussion, specifically regarding the apparent effects of increased urban land cover and decreased riparian forested areas increasing TA concentrations at the sub-watershed pour point. The results of the TA model are similar to TP for urbanization in magnitude and direction of the correlations. The results of the TA model are similar in sign to TP correlations with percent riparian forest, but the correlations are much weaker for TA than for TP (Table 5.3). The TA model, as with the TP model, indicates that the concentration of TA at the sub-watershed pour point is ameliorated when urban land is at greater distances from streams and when forested areas are more prevalent in the sub-watershed, particularly when those forested areas are close to streams. This may be because there are greater opportunities for soil/litter accumulation and transformation in forested areas between TA sources and streams compared with human-built areas (i.e., paved areas, non-vegetated areas) where TA accumulation and transformation is minimal. The results suggest that the loss of forested areas at the riparian and sub-watershed scales, along with replacement of these forested areas with urban land cover (including the associated activities of inadequately treated municipal wastewater and storm water runoff), are the dominant causes of increased TA in streams of the Ozarks.

5.1.4.3 Escherichia coli *Model*

The *E. coli* PLS model results are somewhat different from the TP and the TA model results but share some important similarities: The density of streams within a sub-watershed may have a mitigating effect on *E. coli* cell counts in streams of a sub-watershed, as with TP; and percent urban land cover (and associated road metrics) within the full sub-watershed, and within the riparian areas of a sub-watershed, respectively, are progressively more correlated (as a function of distance from streams) with the *E. coli* cell counts in surface water, as was the case with the TP and the TA models. The *E. coli* model results suggest that an increased percentage of forestland cover in areas farther from streams (e.g., Rfor120 and Pfor) increases the *E. coli* cell counts at the sub-watershed pour point but that an increased percentage of land cover in areas closer to streams (e.g., Rfor0 and Rfor30) decreases the *E. coli* cell counts. This may be a result of *E. coli* presence in fecal material of animals in and near forested areas at a broad scale, but may also be an inconclusive result related more to the fact that *E. coli* is a biotic entity that is less reliable as a landscape scale response variable than TP and TA, which are relatively more influenced by the physical and chemical characteristics of soils and hydrology.

Riparian human-use metrics (Rhum) exhibited the same relationships observed for the TP models, except for Rhum120, which may also be confounded by the biotic nature of *E. coli* as a landscape scale response variable. However, the correlations between Rurb0, Rurb30, Rurb120, and Purb (and the relative coefficient magnitudes) suggest that urbanization in the riparian and in the sub-watershed as a whole may increase the cell count of *E. coli* at the sub-watershed pour point. A critical (i.e., highest VIP) but weak positive correlation between elevation and *E. coli* cell count may indicate that sub-watersheds that are dominated by headwater streams are more likely to have greater counts of *E. coli* than sub-watersheds that are dominated by slower-flowing rivers and may create conditions for greater dilution.

5.1.4.4 *Model Integration*

The results of the TP and TA models indicate that the percent of forest in a sub-watershed, and particularly in riparian vegetated zones of up to 120-m, is an important and consistent factor in determining the potential concentration of TP and TA at the sub-watershed pour point. These relationships comport with our understanding of the biophysical processes of runoff, uptake, accumulation, and transformation of P and N in the environment.

The PLS results are also consistent with our current understanding of the biophysical relationships between urbanization and the inputs of P, N, and bacteria to streams. Thus, it is demonstrated that Rurb0, Rurb30, Rurb120, and Purb are the most reliable indicators of TP, TA, and *E. coli* for the Ozarks. Additionally, the results of the TP, TA, and *E. coli* models indicate that the percent of urban land cover in a sub-watershed, and particularly in riparian (formerly) vegetated zones of up to 120 meters, is an important and consistent factor in determining the potential concentration of the respective surface water constituent at the sub-watershed pour point. These relationships are identical in terms of increasingly strong correlations at distances closer to the stream banks and their general magnitudes within a variable (i.e., the range of coefficients for each variable).

Interpreting the integrated parameter PLS results for the Ozarks, it was determined that 20 of the 46 tested broad-scale landscape metrics are predictive of four different sub-watershed vulnerability states, as follows:

1. "Most vulnerable" sub-watersheds (i.e., high TP concentrations, high TA concentrations, and high *E. coli* cell counts)

2. "Highly vulnerable" sub-watersheds (i.e., low TP concentrations, high TA concentrations, and high *E. coli* cell counts)

3. "Moderately vulnerable" sub-watersheds (i.e., moderate TP and TA concentrations and moderate *E. coli* cell counts)

4. "Least vulnerable" sub-watersheds (i.e., low TP and TA concentrations and moderate *E. coli* cell counts)

The vulnerability states demonstrated in Figure 5.6 are generalized simultaneously for the three PLS models in Table 5.3, which are mapped among the 244 sub-watersheds in Figure 5.1.

The results of the TP and TA PLS models indicate that the percent of urban land cover and the percent of forest in a sub-watershed, and particularly in riparian (formerly) vegetated zones of up to 120 meters, are important and consistent factors in determining the potential concentration of a surface water constituent at the sub-watershed pour point. Our results suggest that non-vegetated areas along streams that are correlated with greater TP and TA concentrations are dominated by urban land cover, and those areas that have forest vegetation within 120 meters of streams are responsible for lesser concentrations of TP and TA. However, the loss of forested areas at the riparian and sub-watershed scales is related to the replacement of these forested areas with urban land cover and agricultural land cover where associated activities of inadequately treated municipal wastewater, agricultural runoff, feedlots, and storm water runoff may be sources of TP and TA. Our results indicate that, particularly in the more urbanized sub-watersheds of the Ozarks, implementation of specific forest restoration and runoff reduction/mitigation practices are important components of improving water quality, such as the attenuation of eutrophication:

- Reducing nonpoint source inputs of nutrients (e.g., N and P compounds) to streams.

- Reestablishing trees and bushes along stream banks to reduce incident sunlight and water temperature, thereby reducing aquatic plant growth and photosynthesis, and to trap nutrient runoff.

- Restoring wetlands (where appropriate in the Ozarks, although wetlands are not a dominant feature) and other riparian vegetated areas that will facilitate the physical interception, accumulation, and transformation of nutrients, thus reducing plant growth, photosynthesis, pH levels, and ammonia concentrations in the stream.

Although the *E. coli* model has some important similarities to the TP and TA models, the *E. coli* model is less clear in terms of the ecological mechanisms and the implications for management. Also, riparian or sub-watershed restoration practices, related to the urban and forest metrics discussed, are recommended in this landscape so as to limit inputs from wastewater treatment outfalls; nonpoint source agricultural field runoff; urban runoff; leaks from septic systems associated with human habitation; and locations that contain cattle, poultry, or swine wastes.

The specific actions suggested in this chapter, based on the project results, relate to the discussed landscape metrics, which are correlated with and indicate the likely water chemistry of each sub-watershed as measured at its pour point. The Ozarks case study results demonstrate that an integration of landscape metrics of land cover type (e.g., forest), land cover quantity (e.g., percentage of forest), land cover location (e.g., riparian zones), and land cover configuration (e.g., the pattern of forest patches), combined with surface water data, can be used as sub-watershed scale indicators of water quality condition. Additional analyses are being conducted using temporal data to explore the effect of short-term and long-term change on these and other water quality parameters.

The results of PLS analyses provide watershed managers with the first broad-scale predictions that can be used to explain how land cover type, quantity, location, and configuration may affect the chemical and biological characteristics of surface water in the Ozark's White River. The landscape factors that were used to conduct the PLS analyses are currently being used by USEPA Region 7 (Kansas City, Kansas) and members of The Upper White River Foundation to determine the vulnerability of surface water resources in the region, to assess sub-watershed condition, and to plan restoration efforts at regional and local scales (Lopez, in preparation).

The critically important work described in this case study provides the current and sound data necessary for measuring watershed health and function, and provides an innovative example that can be modified by those wishing to quantitatively utilize remotes sensing data for the purpose of measuring water quality. Key learnings from this chapter should be that, although there is often a limited amount of field water quality data available for landscape scale projects, there are great opportunities to utilize advanced geospatial and statistical techniques to determine the correlations between landscape condition and the water (bio)chemistry of the same landscape. These approaches allow us to look to the future uses of remote sensing for landscape ecology in a new way, and clearing a path forward to meet the growing needs for meaningful landscape science that connects with user communities. The theme explored in this case study, that is, of increased specificity of the ecological factors on the ground, helps us increase the focus on those issues of importance to regional and local decision makers, which certainly carries forward into the next and last case study of this book. Specifically, the next case study focuses on detailed analyses of remote sensing and field-based information, in combination with a synoptic view of the landscape, to provide new ways to view, analyze, portray, and solve environmental challenges that impact both ecology and society, at a regional to national scale.

5.2 Case Study: Watersheds of the Missouri River and the Mississippi River

The previous project utilizes remote sensing to specifically develop landscape-based predictive models of watershed conditions, specifically in terms of water quality, and the relationships of that water quality and the gradients of condition among multiple watersheds where the complexity of analysis and interpretation is high, with solutions for overcoming these complexities. In the following project, we continue the approaches of overcoming such complexities by leveraging field-based information so as to meet the priority needs of local to regional planners, specifically as it pertains to the health, safety, and well-being of humans on the landscape.

A key element of both the previous and the following project is to provide decision makers with information about the benefits of including ecosystem components, like wetlands within a landscape, rather than removing or degrading them, and how such inclusion can ameliorate flooding risks downstream in large river ecosystems, all through the quantitative use of remote sensing and landscape ecology methodologies. The result of this approach is an improvement in society, through the protection and enhancement of the lives of people who live within the landscape matrix.

Key to these analyses is an understanding of the entirety of the landscape context. Throughout the Lower Missouri River and Upper Mississippi River Watersheds, much of the landscape near the rivers has been heavily converted to row crop agriculture, pasture, or areas of human habitation or commerce, since settlement by Europeans occurred, as far back in history as the turn of the eighteenth century. The increased presence of these new areas of agriculture or urban land, especially in the nineteenth and twentieth centuries are thought to have cumulatively contributed to the overall degradation of the biophysical processes of adjacent and downstream areas along these river systems, and may be a major driver of hydrologic and physical changes that occur along the river systems, such as erosion, lowered water quality, and loss of animal and plant habitat. The loss of ecosystem services along these river courses and, importantly, within the larger watershed basin of these "great rivers" of America can be effectively addressed at a variety of scales, utilizing remote sensing data and landscape ecological methodologies in concert, to determine ecosystem dynamics as well as to provide planners with excellent information about mitigation and adaptation strategies for improving the future conditions of the basin. It is important to note that the methodologies utilized in this case study help to inform critical safety measures to protect life and property throughout the river basins, in addition to guiding important ecological outcomes.

Due to both the immensity of the areas involved and the complexity of the societal and ecological issues involved, this case study specifically takes on filling the gaps in knowledge regarding the interactions of human development of the landscape within watersheds of the tributaries that feed the Missouri River and Mississippi River. The work in this case study interweaves the evaluation of variability of landscape conditions among tributary watersheds and the potential for influence of landscape characteristics among those watersheds, with a special focus on river hydrodynamics within watersheds that are directly driven by landscape conditions. The work in this section depends heavily on several landscape ecological modeling approaches that depend upon a variety of biophysical information that is obtained through both remote sensing and field-based data collection.

The landscape ecological approaches undertaken in this section reveal the potential impact of land cover configuration on the physical characteristics of riverine and riparian areas of 22 tributaries to the Lower Missouri River and the Upper Mississippi River (Figure 5.7), and assesses the potential for the contribution of riparian vegetated areas to a particular watershed's physical conditions. In order to analyze the river hydrodynamics, we integrated the use of remote sensing, GIS data derived from remote sensing, existing field data, and a priori knowledge of the basin's ecology, that is, the interrelationships between biotic and abiotic elements of the landscape.

The several outputs of this evaluation of the Missouri River and Mississippi River Basins (i.e., the tributaries to these large rivers) yield an important opportunity to provide regional decision support information for planning, as well as methods for continuing the monitoring and adjustment of future scenarios, which is an important feature of these models, particularly in the face of the environmental uncertainty that accompanies climate change.

The variety of river/riparian models in this project help to support the critical societal elements (i.e., land use planners, conservation organizations, and citizens) within the basin, to determine the relative influence of land cover characteristics on river flow and the influence on riparian ecosystems. The results or methods of the finer-scale hydrologic models for the Kansas River are intended to demonstrate how they can be applied to each of the 22 tributaries outlined in the broader-scale analyses, listed and summarized in Figure 5.7, and as part of a methodology that broadly analyzes a large set of watersheds, with the intent to specifically target sub-watersheds or

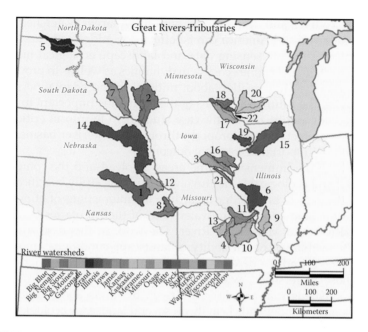

FIGURE 5.7
The 22 tributary river basins analyzed in the Missouri River and Mississippi River Basin project: **(1) The Big Blue River** (http://cfpub.epa.gov/surf/huc.cfm?huc_code=10270205, checked January 25, 2017), **(2) the Big Sioux River** (https://cfpub.epa.gov/surf/huc _code=10170203, checked January 25, 2017), **(3) the Des Moines River** (https://cfpub.epa .gov/surf/huc.cfm?huc_code=07100009, checked January 25, 2017), **(4) the Gasconade River** (https://cfpub.epa.gov/surf/huc.cfm?huc_code=10290203, checked January 25, 2017), **(5) the Grand River** (https://cfpub.epa.gov/surf/huc.cfm?huc_code=10130303, checked January 25, 2017), **(6) the Illinois River** (https://cfpub.epa.gov/surf/huc.cfm?huc_code=07130011, checked January 25, 2017), **(7) the James River** (https://cfpub.epa.gov/surf/huc _code=10160011, checked January 25, 2017), **(8) the Kansas River** (https://cfpub.epa.gov/surf /huc.cfm?huc_code=10270104, checked January 25, 2017), **(9) the Kaskaskia River** (https:// cfpub.epa.gov/surf/huc.cfm?huc_code=07140204, checked January 25, 2017), **(10) the Meramec River** (https://cfpub.epa.gov/surf/huc.cfm?huc_code=07140102, checked January 25, 2017), **(11) the Missouri River** (https://cfpub.epa.gov/surf/huc.cfm?huc_code=10300200, checked January 25, 2017), **(12) the Big Nemaha River** (https://cfpub.epa.gov/surf/huc.cfm?huc _code=10240008, checked January 25, 2017), **(13) the Osage River** (https://cfpub.epa.gov /surf/huc.cfm?huc_code=10290111, checked January 25, 2017), **(14) the Platte River** (https:// cfpub.epa.gov/surf/huc.cfm?huc_code=10200202, checked January 25, 2017), **(15) the Rock River** (https://cfpub.epa.gov/surf/huc.cfm?huc_code=07090005, checked January 25, 2017), **(16) the Skunk River** (https://cfpub.epa.gov/surf/huc.cfm?huc_code=07080107, checked January 25, 2017), **(17) the Turkey River** (https://cfpub.epa.gov/surf/huc.cfm?huc_code=07060004, checked January 25, 2017), **(18) the Iowa River** (https://cfpub.epa.gov/surf/huc.cfm?huc _code=07060002, checked January 25, 2017), **(19) the Wapsipinicon River** (https://cfpub.epa .gov/surf/huc.cfm?huc_code=07080103, checked January 25, 2017), **(20) the Wisconsin River** (https://cfpub.epa.gov/surf/huc.cfm?huc_code=07070005, checked January 25, 2017), **(21) the Wyaconda River** (https://cfpub.epa.gov/surf/huc.cfm?huc_code=07110001, checked January 25, 2017), and **(22) the Yellow River** (https://cfpub.epa.gov/surf/huc.cfm?huc_code=07060001, checked January 25, 2017).

portions thereof for restoration, in support of ongoing conservation efforts within the Missouri River and Mississippi River tributary watersheds.

5.2.1 Landscape Metrics among Great Rivers Tributary Basins

A fairly recent environmental management trend is the move away from simple, local-scale assessments and a move toward the more complex, multiple-stressor regional types of assessments. Landscape ecology principles outlined in Chapter 2 provide the theory behind these assessments, while remote sensing theory and practice, as well as geospatial analysis theory and GIS technology, supplies the key tools to implement them. As in the Great Lakes case study (Chapter 3 and Chapter 4), a sensible application of GIS for exploring the computation of landscape metrics across a vast landscape is the most prudent first step to quantify the environmental conditions or potential vulnerability of a spatial reporting unit, and among reporting units across vast areas of the landscape (e.g., watersheds, HUCs). The landscape metrics in the Missouri River and Mississippi River case study have accordingly been divided into five groups, based upon hydrologic basin analysis units of the larger tributary basin, among the 22, by the following categories, with a single key example of each metric type provided as a figure:

1. Landscape characteristics (Figure 5.8)
2. Riparian characteristics (Figure 5.9)
3. Human stressors (Figure 5.10)
4. Physical characteristics (Figure 5.11)
5. Soil/landform (Figure 5.12).

5.2.2 Flooding Futures among Tributary Basins

Once the landscape characteristics are assessed and quantified, as in Section 5.2.1, a number of modeling techniques can be used to demonstrate flood potential within the landscape, utilizing remote sensing and GIS data. Figure 5.13 to Figure 5.18 demonstrate such an approach, depicting the key impacts of flooding scenarios on three important land cover types: urban areas, forested areas, and wetland areas. Although the flood effects can be assessed far into the expansive floodplains of these tributary basins, which may be desirable depending upon the decision maker's questions and needs, we have demonstrated one such approach, which focuses primarily upon the riparian zone of streams in a watershed. Customarily, the riparian area of streams is focused on the 80–120-m "buffer" area on the bank of a stream or river. Considering the resolution of the data utilized in this project and the scale at which our assessment is focused (large areas among a large number of tributary watersheds), we demonstrate the flooding impacts among the tributary basins by analyzing 120-m riparian buffer areas, adjacent to

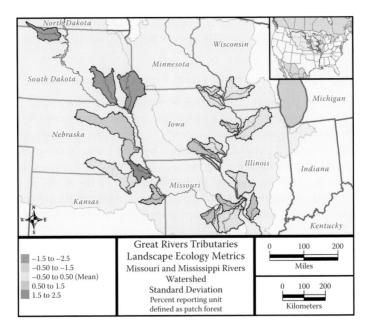

FIGURE 5.8

Landscape characteristics are initially assessed for the Great Rivers Tributaries study area, including such metrics as *percent patch forest*. Percent patch forest is calculated using a moving 270-meter-square window (9 pixels × 9 pixels) across the land cover. When the percent of forest in the window is less than 40%, the forest cell in the center of the window is classified as *patch*. The number of patch forest cells in the reporting unit is then divided by the reporting unit's total land area (the total number cells in the watershed less those cells classified as water). Percent patch forest is less likely to be connected or to provide interior habitat for some organisms. Standard deviation is utilized in metric analyses to show the amount a feature's attribute value that varies from the mean value of the total distribution. Selected class breaks are generated by successively adding or subtracting the standard deviation from the mean. A two-color ramp is used to emphasize values above or below the mean. (Data from the National Land Cover Database.)

streams and rivers. This value can easily be adjusted as needed, however comports with the typical zones that are analyzed within the literature. Initial analyses in this case study provide a synoptic look at the spatial differences among all 22 tributary watersheds for two flood extents (base to medium flooding, and base to maximum flooding). The flood extent difference represents the effect of the flooding on areas within the 120-m buffer zone from the main stem. As with the metrics in the prior case study, proportions and patterns among the basins is determined by the input variables, derived from a number of remote sensing and GIS data sets, as summarized for each model.

The initial synoptic flood modeling results in this case study are a result of outputs from Vflo flood modeling software (http://www.vieuxinc.com/vflo.html, checked January 25, 2017), which utilizes custom settings (i.e., rainfall amount,

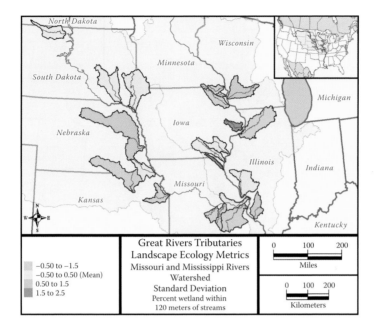

FIGURE 5.9
Riparian characteristics assessed for the Great Rivers Tributaries study area includes the percentage of wetland land cover within 120 meters of a stream, which is calculated by summing the total number of wetland land cover cells underneath stream segments in the reporting unit and within a three cell buffer (120 meters) and dividing by the stream corridor's total land area (all cells 120 meters adjacent to streams minus those classified as water). Cells inside the buffer zone but outside of the reporting unit boundary are ignored. Forests and wetlands within riparian areas can act to attenuate flood waters and runoff as well as remove pollutants from the water column. Forests and wetlands in the riparian zone can also provide habitat for a wide variety of plant and wildlife species that provide important ecological functions in the landscape. Standard deviation is utilized in metric analyses to show the amount a feature's attribute value that varies from the mean value of the total distribution. Selected class breaks are generated by successively adding or subtracting the standard deviation from the mean. A two-color ramp is used to emphasize values above or below the mean. (Data from the National Land Cover Database and National Hydrography Dataset Plus.)

duration, flood type/intensity) in each of the 22 tributary watersheds, for each of the two hydrologic scenarios previously described. Through iterative analyses, flood extents that fell within the study area's 120-m stream/river main stem buffer were selected and displayed, among the 22 tributary watersheds for comparison and analyses. It is important to recognize that the Vflo-simulated flood extent in these initial analyses are the product of remote sensing derived GIS information, without the further benefit of field-based information, and thus depict more general effects of hydrodynamic and watershed land cover change than the focal analyses of the Kansas River Watershed, described in the next section of this chapter. This is not to say that these initial analyses

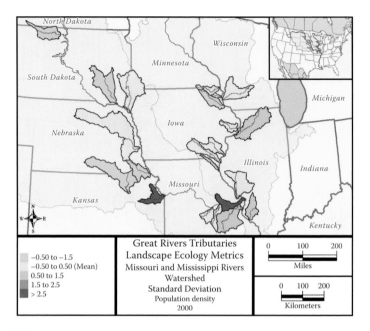

FIGURE 5.10
Human stressors assessed for the Great Rivers Tributaries study area includes human population density of the landscape. Population density was calculated by summing number of people living in the reporting unit and dividing by the reporting unit area. Where census units were not completely contained within the watershed reporting unit, population was apportioned by area. High population densities are generally well correlated with high amounts of human land uses, especially urban and residential development. Large areas of development often involve substantial modification of natural vegetation cover and may have significant effects on wildlife habitat, soil erosion, and water quality. Standard deviation is utilized in metric analyses to show the amount a feature's attribute value that varies from the mean value of the total distribution. Selected class breaks are generated by successively adding or subtracting the standard deviation from the mean. A two-color ramp is used to emphasize values above or below the mean. (Data from the Census Block Groups.)

are without value at all, but rather to demonstrate what we understand from earlier case studies, that both broad-scale and finer-scale data sets and analyses provide improved analytical capabilities, and their complementarity provides power to remote sensing–based landscape ecological projects.

5.2.3 Hydrologic Change Analysis in the Kansas River Watershed

Broad-scale analyses in the prior section suggests that a number of the tributary watersheds are vulnerable to hydrologic change, and that the influence of land cover change is substantial. For this reason, a detailed analysis of each tributary was warranted, and among them we have focused on the Kansas River Watershed (Figure 5.19) for additional insight as an example of analyzing the landscape ecological impacts of predicted watershed change,

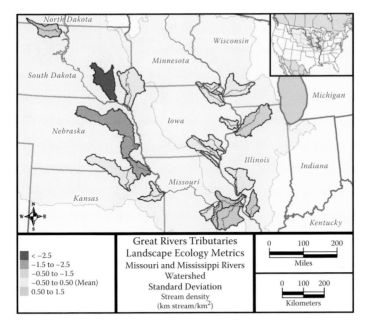

FIGURE 5.11
Physical characteristics assessed for the Great Rivers Tributaries study area includes stream density (km stream/km²). Stream density was calculated by summing the total length of streams in the watershed reporting unit and dividing by the total watershed area. (Data from National Hydrography Dataset Plus.)

with regard to hydrology. All of the watersheds in the prior section are subject to the same analyses that are demonstrated here, in the Kansas River Watershed, which provides an ideally representative landscape among the 22 tributary watersheds in the Lower Missouri River and Upper Mississippi River basins.

The Kansas River is located in northeastern Kansas (Figure 5.7). It is formed at the confluence of the Republican River and the Smoky Hill River, just east of Junction City, Kansas. From the confluence of these two rivers, the Kansas River flows 274 km to Kansas City, where it discharges to the Missouri River. The Kansas River Valley is approximately 222 km long with the surplus length a result of the river's meandering across the valley. The Kansas River drops approximately 98 meters from its starting point along the length of the river valley, with a slope of less than 0.6 m per km. The river valley averages 4.2 km in width and the widest stretch of the river valley is between Wamego and Rossville, where it is approximately 6.4 km wide. Below Eudora, the valley narrows to less than half the maximum width to 2.4 km—in parts of this river reach the valley is 1.6 km or less in width (KGS 1998). The total river basin area is approximately 14,200 km². Land use throughout the study area is predominantly agricultural, with some urban areas, and includes the towns of Junction City, Manhattan, Topeka, Lawrence, and Kansas City.

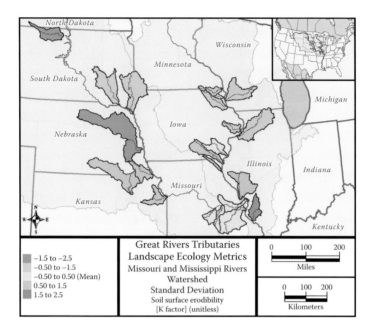

FIGURE 5.12
Soil/landform assessed for the Great Rivers Tributaries study area includes soil surface erodibility (K factor) among the watersheds. RUSLE weighted-average effect of inherent soil surface erodibility (K factor), which is from STATSGO data, is computed on a cell-by-cell area basis. An increase in soil erodibility may indicate an increase in the amount of runoff of sediment and chemical constituents associated with sediment (e.g., phosphorus) to streams and lakes. (Data from National Elevation Dataset (30 meters) and STATSGO Soils.)

Counties through which the Kansas River flows support a growing population of nearly 800,000 people (USGS 2005). The Kansas River drains the area defined by four 8-digit Hydrologic Unit Code (HUC) cataloging units: HUC 10270101 (Upper Kansas Watershed), HUC 10270102 (Middle Kansas Watershed), HUC 10270103 (Delaware River Watershed), and HUC 10270104 (Lower Kansas Watershed).

Flooding is a major natural hazard that impacts different regions across the world every year, and this watershed is no exception. The use of remote sensing to assess the human safety aspects of this issue at the landscape level is well established (Patel and Srivastava 2013). Recently in the United States, both the Missouri River and Mississippi River regions have experienced unprecedented flooding events, which have caused fatalities, evacuations, and large financial losses. In addition, climate change is expected to continue the intensity and extremity of the risks associated with storm events (Milly et al. 2002) and a concomitant increase in the frequency of flash floods. Accordingly, large landscape scale floods in many regions are very likely to occur and perhaps increase into the future (Alley et al. 2003; Parry et al. 2007). Moreover, the world is currently undergoing the largest rate of

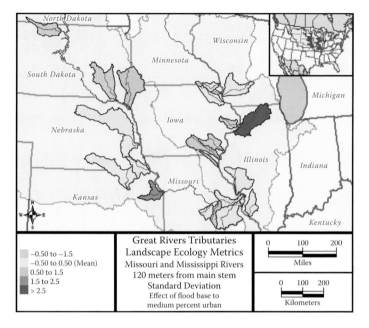

FIGURE 5.13
Percent urbanized areas assessed for the Great Rivers Tributaries study area under "Base to Medium" flow conditions. The percentage of urban land use is calculated by dividing the number of urban land cover cells in the riparian zone (i.e., within 120 meters of a stream) for the entire watershed reporting unit by the total number of cells in the same reporting unit boundary, less those cells classified as water (total land area). High amounts of urban land indicate substantial modification of natural vegetation cover and may have profound effects on wildlife habitat, soil erosion, and water quality; they are also highly susceptible to flood events in these vulnerable areas of the landscape. (Data from the National Land Cover Database, National Elevation Dataset Plus, and Census Block Groups.)

urbanization in history. In 2008, for the first time in history, approximately half of the world's population resided in urban areas and these residents are consequently expected to experience the majority of the future global population growth on our planet (UN 2010). From 1982 to 1997, the amount of land devoted to urban and other human-built land uses in the United States increased by 34% (USDA 2001). Urbanization, which results in an increase in impervious pavement and other surfaces, generally increases the size and frequency of floods in the vicinity and downslope/stream from development, and may expose communities to increasing flood hazards (Parker 2000; USGS 2003), resulting in an increasing focus by planners and land managers on the role of urbanization in the prediction of flood levels and damage, primarily for disaster management and urban and regional planning (ASME 2008; Milly et al. 2008).

Areas where flooding and human conversion/occupation of the land is high, the intial analyses of the 22 tributaries to the Lower Missouri River

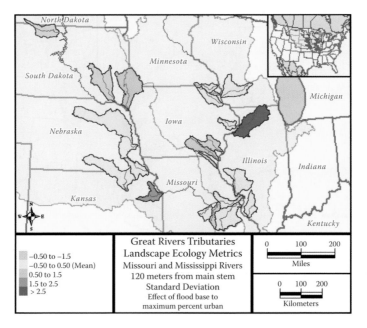

FIGURE 5.14
Percent urbanized areas assessed for the Great Rivers Tributaries study area under "Base to Maximum" flow conditions. The percentage of urban land use is calculated by dividing the number of urban land cover cells in the riparian zone (i.e., within 120 meters of a stream) for the entire watershed reporting unit by the total number of cells in the same reporting unit boundary, less those cells classified as water (total land area). High amounts of urban land indicate substantial modification of natural vegetation cover and may have profound effects on wildlife habitat, soil erosion, and water quality; they are also highly susceptible to flood events in these vulnerable areas of the landscape. (Data from the National Land Cover Database, National Elevation Dataset Plus, and Census Block Groups.)

and Upper Mississippi River, as well as a history of the confluence of these two phenomena led us to select the Kansas River for further landscape ecological analyses. The landscape in Kansas (among the top ten states for annual flood loss in the United States between 1972 and 2006 [Changnon 2008]) follows the global pattern described earlier, where increasing population growth and urban development, higher likelihood of exposure to flood damage, and agricultural losses from flooding is likely to increase in the future. The main river ecosystem, the Kansas River, has also been prone to flooding over the last century. For example, the Kansas River region has experienced two significant flood events, one in 1951 and another in 1993. The damages from the 1951 flood were exceptional, with 19 people killed and approximately 1,100 people injured, with total financial damage estimates as high as $2.5 billion (Juracek et al. 2001). During the height of the July 13, 1951, flood, nearly 90% of the flow in the Missouri River at Kansas City came from

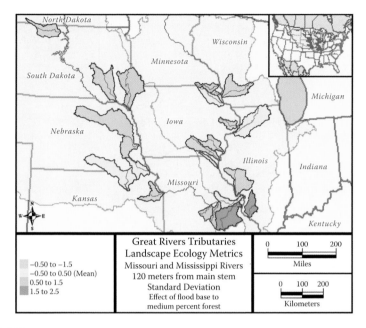

FIGURE 5.15
Percent forested areas assessed for the Great Rivers Tributaries study area under "Base to Medium" flow conditions. The percentage of forested land cover is calculated by dividing the number of forestland cover cells in the riparian zone (i.e., within 120 meters of a stream) for the entire watershed reporting unit by the total number of cells in the same reporting unit boundary, less those cells classified as water (total land area). Forested areas provide critical habitat for a wide variety of valuable plant and wildlife species, and provide important ecosystem services to those communities and people living in and downstream. (Data from the National Land Cover Database and National Elevation Dataset Plus.)

the Kansas River, tributary to the Missouri River that comprises approximately 12% of the Missouri River's drainage basin (Juracek et al. 2001). The other historic flood of note in the larger landscape, including the Kansas River Watershed, was in 1993, which affected nine U.S. states and resulted in approximately $15 billion in flood damages. From July 22nd to July 24th, 1993, 50–330 mm of rain fell throughout Kansas and Nebraska, which contributed an enormous input of runoff to an already-saturated landscape and full reservoirs throughout the Kansas River basin. Eighteen of the 163 USGS stream gauges in operation in Kansas during 1993 measured record maximum peak daily flows and 69 stations measured the highest mean annual streamflow during their period of record for Water Year 1993 (Combs and Perry 2003).

Among the many applications of remote sensing, one of the key uses is the production of associated GIS data sets and landscape scale flood models. Calculating risks from floods with these models is an effective approach

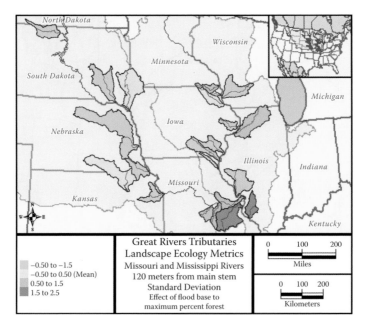

FIGURE 5.16
Percent forested areas assessed for the Great Rivers Tributaries study area under "Base to Maximum" flow conditions. The percentage of forested land cover is calculated by dividing the number of forestland cover cells in the riparian zone (i.e., within 120 meters of a stream) for the entire watershed reporting unit by the total number of cells in the same reporting unit boundary, less those cells classified as water (total land area). Forested areas provide critical habitat for a wide variety of valuable plant and wildlife species, and provide important eco-system services to those communities and people living in and downstream, such as flood attenuation, sediment load reduction, aesthetics, and recreational/educational benefits. (Data from the National Land Cover Database and National Elevation Dataset Plus.)

for assessing potential for harm to the people living within landscapes as well as their property, as well as ecosystem changes. These important models demonstrate in quantitative terms how prone to flooding certain areas of the landscape are, and the return periods of such flood events. To mitigate flood risk, reservoirs and levees have been historically utilized and in recent years the inclusion of restored or created wetlands in the landscape have also been studied for their placement in the landscape for the purposes of flood attenuation and water quality improvement (e.g., Mitsch et al. 2001; Crumpton et al. 2006). The growing recognition and appreciation of wetlands' natural capability for short-term surface water storage, and capability for reducing downstream flood peaks, has led to an increase in focus on methods to mitigate flooding events at a broad landscape scale, primarily by attenuating the flood pulse that moves downstream after large precipitation events, through riparian wetland restoration and conservation. Wetlands

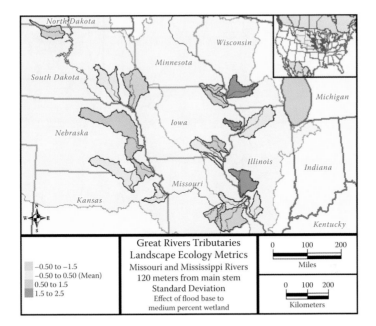

FIGURE 5.17

Percent wetland areas assessed for the Great Rivers Tributaries study area under "Base to Medium" flow conditions. The percentage of wetland cover is calculated by dividing the number of wetland land cover cells in the riparian zone (i.e., within 120 meters of a stream) for the entire watershed reporting unit by the total number of cells in the same reporting unit boundary, less those cells classified as water (total wetland area). Wetlands provide critical habitat for a wide variety of valuable plant and wildlife species, and also provide important ecosystem services to those communities and people living in and downstream, such as flood attenuation, sediment load reduction, aesthetics, and recreational/educational benefits. (Data from the National Land Cover Database and National Elevation Dataset Plus.)

also benefit the biological, wildlife habitat, and water quality characteristics of the large riverine and riparian ecosystems, which are often diffuse and numerous throughout the landscape (Lewis 1995; Hey and Phillippi 1995; Wamsley et al. 2010). Wetlands' tremendous capability for short-term surface water storage, and therefore reduction in downstream flood peak pulses from large precipitation events, has made them a very important focus for a variety of hydrologic studies, including landscape ecological analyses.

The landscape ecological approach described in this (Kansas River) focal portion of the case study involves the development of predictive models that describe the surface water storage capability of wetlands within the larger riverine ecosystems of large landscapes, utilizing a specific set of tools that is dependent upon both remote sensing data and knowledge of the physical characteristics of elements in the landscape, from both the ecological

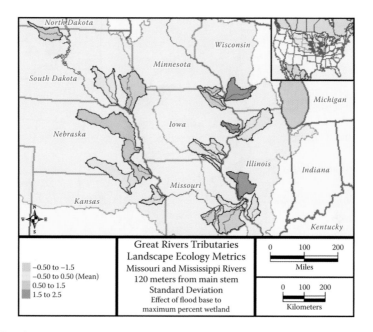

FIGURE 5.18
Percent wetland areas assessed for the Great Rivers Tributaries study area under "Base to Maximum" flow conditions. The percentage of wetland cover is calculated by dividing the number of wetland land cover cells in the riparian zone (i.e., within 120 meters of a stream) for the entire watershed reporting unit by the total number of cells in the same reporting unit boundary, less those cells classified as water (total wetland area). Wetlands provide critical habitat for a wide variety of valuable plant and wildlife species, and also provide important ecosystem services to those communities and people living in and downstream, such as flood attenuation, sediment load reduction, aesthetics, and recreational/educational benefits. (Data from the National Land Cover Database and National Elevation Dataset Plus.)

and hydrological perspectives, as developed by the Hydrologic Engineering Center (HEC), an organization within the U.S. Army Corps of Engineers' Institute for Water Resources. The HEC developed several landscape-based tools that allow for riverine and riparian modeling, which include the Hydrologic Modeling System (HEC–HMS), to build a hydrologic model for a river, while the River Analysis System (HEC–RAS) is used to build the hydraulic model for the river. These two tools are commonly used and have been employed for:

- Building flood forecasting and flood inundation models (Knebl et al. 2005; Whiteaker et al. 2006)
- Analyzing different flood control alternatives (Benavides et al. 2001)
- Addressing social impacts of small dam removals (Wyrick et al. 2009)
- Developing a flood early warning system (Matkan et al. 2009)

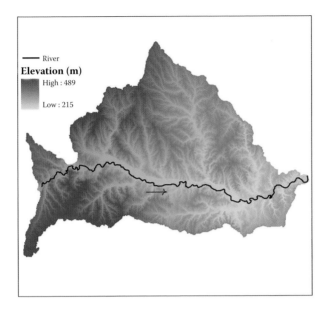

FIGURE 5.19
Kansas River drainage area, including elevational gradients, in northeastern Kansas, United States. Arrow indicates Kansas River flow direction, from west to east.

The primary objectives of this focal portion of the case study are to:

- Evaluate the impacts of future land use and land cover change, in the backdrop of 100-year design storms (considered to be an extreme storm event), on the peak runoff and flood inundation extents for the Kansas River
- Evaluate the potential benefit of wetlands for flood attenuation, specifically in the vicinity of the Kansas River

The approaches used in this case study can also be applied to any riverine/riparian ecosystem, and the surrounding landscape that it is embedded within, around the world (although this application may be potentially limited by data availability) by utilizing analogous methods, remote sensing data, associated GIS data sets, and analytical tools.

5.2.4 Determining Riverine, Riparian, and Floodplain Landscape Conditions

The landscape scale hydrological model for the Kansas River Basin was designed using HEC–HMS and the landscape scale hydraulic model was created using HEC–RAS. USGS stream gauges and National Climatic Data Center (NCDC) weather station data were used to calibrate and validate

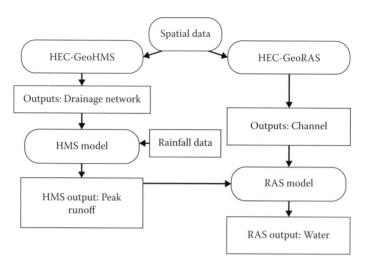

FIGURE 5.20
Conceptual model of the HEC modeling process used in the Kansas River Basin.

models for different historic storm events. Future land use and land cover scenarios were developed in a GIS for the years 2020, 2030, and 2040 with increasing levels of urbanization as time progresses in the models. To evaluate the impact of land use and land cover changes, the hydrologic model was used to generate runoff estimates for the SCS 100-year 24-hour design storms. Then, using the hydraulic model for those runoff estimates, the water levels and flooding extents were generated. Finally, the wetland potential for flood mitigation/attenuation was evaluated. Figure 5.20 outlines a conceptual model, which outlines the process of each step, from spatial data preparation and input to the HEC–RAS output stage.

5.2.4.1 The Hydrologic Model

The HMS simulates the precipitation-runoff processes within the larger landscape by simulating watershed processes; HMS is a generalized modeling system capable of representing many different watersheds. The model uses a deterministic mathematical modeling approach to compute various components of the hydrologic cycle. Hydrographs produced by HMS were used directly or in conjunction with other software for studies of water availability, urban drainage, flow forecasting, future urbanization impact, reservoir spillway design, flood damage reduction, floodplain regulation, and systems operation. A model of the watershed was then constructed by separating the hydrologic cycle into manageable components and then constructing boundaries around the watershed of interest (HEC 2010a). For developing the watershed properties, topographic data was utilized by developing a digital elevation model of the study area; we used a USGS DEM to develop the watershed properties using another of the HEC tools, the GIS-based HEC–GeoHMS. HEC–GeoHMS allows the user to visualize spatial information, document watershed characteristics, perform spatial analyses, delineate sub-basins and streams, construct inputs to hydrologic models, and provide output reports. HEC–GeoHMS creates different model files that can be used to develop a hydrological model by HEC–HMS (HEC 2010b). The HMS model requires three main input process parameters:

1. Precipitation loss for overland flow
2. Transformation of overland flow into surface runoff
3. Streamflow routing

The precipitation loss method for overland flow accounts for the infiltration losses; there are multiple methods possible for this approach, all within HMS, and the SCS curve number method (HEC 2010c) was selected, where values are computed from a curve number grid. Curve numbers represent the runoff potential of an area with values ranging from 0 to 100; greater curve number values indicate greater runoff potential. The process for the curve number grid preparation in this project utilizes the National Land Cover Database (http://www.mrlc.gov/nlcd2011.php; checked January 25, 2017) and the Soil Survey Geographic (SSURGO) database (https://www.nrcs .usda.gov/wps/portal/nrcs/detail/soils/survey/?cid=nrcs142p2_053627; checked January 25, 2017). The curve number values utilized for the different land use and land cover classes in this project are from McEnroe and Gonzalez (2003).

The Soil Conservation Service (SCS) unit hydrograph method was utilized for transforming overland flow into surface runoff. In the SCS method, 37.5% of the runoff volume occurs before the peak flow, and the lag time can be approximated by taking 60% of the time of concentration. The lag time used was the length of time between the centroid of rainfall excess and the peak flow of the resulting hydrograph; values were computed using the curve number Lag Time function within GeoHMS (HEC 2010c).

Once excess precipitation has been transformed into overland runoff and routed to the outlet of a sub-watershed, it enters the stream in the model at that point and is added to streamflow routed from upstream; the Muskingum method was selected for this process. The Muskingum method has the parameters K, X, and number of sub-reaches (n), which need to be specifically identified. Muskingum K is essentially the travel time of water flowing through the reach. Muskingum X is the weighting between inflow and outflow influence, which ranges from 0 to 0.5. The number of sub-reaches along the river affects flow attenuation, where one sub-reach gives more attenuation, and increasing the number of sub-reaches decreases the attenuation. K and n values are estimated by using the method in Olivera and Maidment (2000).

Calibration of this hydrologic model for such a large landscape was relatively difficult due to a lack of complete knowledge of antecedent conditions, such as soil moisture content and the level of water in wetlands. Consequently, calibrating a major flood event does not guarantee that the model will accurately simulate another flood event even when they are of the same magnitude (Bengtson and Padmanabhan 1999). Accordingly, multiple peak flow events were selected for calibration. The USGS gauge on the most downstream portion of the river (Gauge 06892350 Kansas River at Desoto, Kansas) was selected for calibrating the model. Hourly runoff data was collected from the USGS Instantaneous Data Archive, which has periodic stream flow records available since 1991. Hourly precipitation data was collected from the NCDC weather stations.

5.2.4.2 *The Hydraulic Model*

Hydrologic Engineering Centers' River Analysis System (HEC–RAS) is a one-dimensional model, intended for hydraulic analysis of river channels. The model is comprised of a graphical user interface, separate hydraulic analysis components, data storage and management capabilities, graphics, and reporting capabilities. Outputs from RAS include water surface elevations (i.e., water levels) and hydraulic properties (e.g., velocity, flow area, and visualization of stream flow) showing extent of flooding. RAS applications include floodplain management studies; bridge and culvert analysis and design; and channel modification studies. The main inputs include the channel and floodplain geometric data, for example, elevation and width (HEC 2010d). The DEM was used to prepare geometric data for the study site, utilizing a GIS-based tool called HEC–GeoRAS. HEC–GeoRAS is a set of GIS-based utilities for processing geospatial data in ArcGIS using a graphical user interface. The interface aids in the preparation of geometric data for import into HEC–RAS and processes simulation results exported from HEC–RAS. The user creates a series of line themes or layers pertinent to developing geometric data for HEC–RAS. The required themes are the Stream Centerline, Flow Path Centerlines, Main Channel Banks, and Cross Section Cut Lines, referred to as the RAS Themes (HEC 2010e).

To simulate historic peak flow water levels with reasonable accuracy, the hydraulic model for the Kansas River was calibrated and validated with the peak flow and water level data from USGS gauges, utilizing data from five USGS gauges located on the Kansas River (Figure 5.21). They include Gauge 06887500 (Kansas River at Wamego, Kansas), Gauge 06888350 (Kansas River near Belvue, Kansas), Gauge 06889000 (Kansas River at Topeka, Kansas),

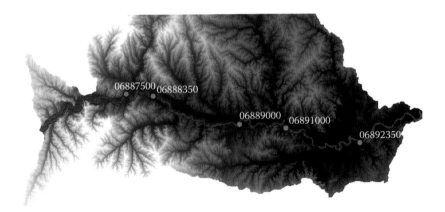

FIGURE 5.21
Location of USGS gauges used for calibration of the hydraulic model for the Kansas River, overlaid on the digital elevation model of the Kansas River Basin.

Gauge 06891000 (Kansas River at Lecompton, Kansas), and Gauge 06892350 (Kansas River at Desoto, Kansas).

5.2.5 Inferring Floodplain Landscape Conditions and Associated River Hydrology

5.2.5.1 Land Use and Land Cover

Landscape scale, future land use, and land cover scenarios in the Kansas River Basin were generated using the Integrated Climate and Land Use Scenarios (ICLUS), a GIS-based tool and data sets for modeling U.S. housing density growth. The output from ICLUS was modified for the purposes of this project to create scenarios representing increasing levels of urban growth and density for the years 2020, 2030, and 2040. ICLUS was developed by the U.S. Environmental Protection Agency's Global Change Research Program at the National Center for Environmental Assessment (ICLUS 2010). ICLUS contains multiple scenarios for housing density and population, from which the model providing the greatest population was selected. ICLUS outputs consist of four land use classes (rural, urban, suburban, and extra-urban); the existing NLCD data were used, with its 15 land use and land cover classes. The ICLUS land use classes: extra-urban, suburban, and urban were assumed equivalent to NLCD urban land use classes 22 (developed low intensity), 23 (developed medium intensity), and 24 (developed high intensity). The ICLUS future output was mosaiced with the present land use and land cover data coverage to create future land use and land cover scenarios for the simulations of change.

TABLE 5.4

Percentage Area Share of the Different National Land Cover Database Classes
for Modeling Scenarios within the Kansas River Basin

Land Use		Scenario (%)			
Class	Description	Baseline	1 (2020)	2 (2030)	3 (2040)
11	Open water	1.7	1.5	1.5	1.5
21	Developed open space	5.0	2.9	2.9	2.8
22	Developed, low intensity	3.0	24.7	2.6	0.9
23	Developed, medium intensity	0.9	2.5	24.9	2.7
24	Developed, high intensity	0.4	0.7	0.8	24.9
31	Barren land, rock, sand, clay	0.2	0.1	0.1	0.1
41	Deciduous forest	2.5	1.4	1.4	1.4
43	Mixed forest	4.3	2.5	2.5	2.5
42	Scrub/shrub	0.8	0.7	0.7	0.7
71	Grasslands, herbaceous	26.8	22.8	22.8	22.8
81	Pasture, hay	25.1	17.7	17.6	17.5
82	Cultivated crops	23.0	17.6	17.5	17.5
90	Woody wetlands	0.2	0.2	0.2	0.2
91	Palustrine forested wetlands	6.2	4.5	4.4	4.4
95	Emergent herbaceous wetlands	0.1	0.1	0.1	0.1

Land use and land cover scenarios were created for simulation (Table 5.4),
including the baseline scenario that uses the NLCD 2001 data, and:

- Scenario 1, that uses the output from ICLUS for the year 2020
- Scenario 2, that interchanges the area for the urban land use classes
  22 (developed, low intensity) and 23 (developed, medium intensity)
  for the ICLUS output in 2030
- Scenario 3, that interchanges the area for the urban land use classes
  22 (developed, low intensity) and 24 (developed, high intensity) for
  the ICLUS output in 2040

By focusing on land use and land cover, different scenarios were created
to reflect increasing urbanization, both in terms of area coverage of development and the density of coverage in the future. ICLUS outputs for 2030 and
2040 show minimal change in urban land use, which has a negligible impact
on future curve numbers, leading to the need for creating more robust
urbanization scenarios as represented by Scenarios 2 and 3.

5.2.5.2 Precipitation

The SCS design storm method was selected for generating future precipitation in the project area, which is a storm simulation method developed

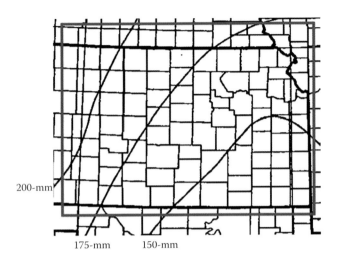

FIGURE 5.22
One-hundred-year return period, 24-hour duration precipitation map for Kansas (modified from Hershfield 1961). The red rectangle highlights Kansas with precipitation contour lines.

by the Natural Resources Conservation Service. Since the 100-year 24-hour storm is considered to have the severest intensity and is typically used in flood modeling studies, it was ideal for the landscape scale simulations in this project to use that model mode. The SCS hypothetical storm method implements four synthetic rainfall distributions from observed precipitation events. Each distribution is 24 hours long and contains rainfall intensities arranged to maximize the peak runoff for a given total storm depth (USDA 1986). Figure 5.22 depicts the different rainfall depth projections crossing Kansas (Hershfield 1961), demonstrating that the 150-, 175-, and 200-mm rainfall isopleths cross the entire state of Kansas. Consequently, all of the rainfall isopleths were selected for analysis.

5.2.5.3 *Wetlands*

Three future scenarios for wetlands were created for modeling the critical effects that wetlands have on floodplain dynamics, especially in terms of flood potential and mitigation/attenuation. In the 2020, 2030, and 2040 scenarios an increasing areal percentage of wetlands within the watershed was modeled. Initial conditions indicated that approximately 5% of the floodplain is wetland (i.e., NLCD value 90, 91, and 95; Table 5.4). Percent wetland area was increased in the study area to 6% in 2020, 8% in 2030, and 10% in 2040, which are all reasonable assumptions based upon the current mind-set of planners and the public; these three scenarios were applied to simulations for the 150-mm storm only. To calculate a flow volume that the wetlands could reasonably retain, a uniform depth of 0.30 meters was assumed for all wetlands, based on literature recommendations (USDA and USEPA 1994; Oregon Department of Environmental Quality 2003).

There is no specific module or component for representing wetlands in HMS, so it was necessary to simulate wetlands in the model by representing them as a "reservoir" at the terminus of a sub-watershed, making the assumption that all flow in a sub-watershed will be intercepted by the reservoir. Wetlands at different locations in the basin were therefore aggregated to simplify their representation in the model, to approximate the hydrologic functions of physically distributed wetlands within the landscape as a whole. Modeling a reservoir also requires making some assumptions, as the appropriate data for creating a storage–discharge relationship and other factors required, because adequately modeling a reservoir in HMS are not currently available. Despite these circumstances, wetlands were successfully modeled using the diversion component in HMS (Bengtson and Padmanabhan 1999). By utilizing a hydrologic diversion in the model, the user is able to specify portions of the flow to be diverted and taken out of the system. Diversions allow for considerable flexibility in setting the rate of diverted flow to incoming flow and various rates can be modeled. Fixing diversions as a rate of incoming flow can mirror the sub-watershed area that contributes to the wetlands. The timing of diversions can be controlled by not allowing water to be diverted until a specific flow is reached. There are two options for diverting water:

1. Flow can start being diverted as soon as runoff is generated.
2. Flow can be diverted after a flow magnitude is exceeded.

The second option is hypothetical and it requires the assumption that the flows to wetlands are controlled through some structural adjustments in the watershed. The volume diverted in both cases would be the same but changing the timing of the diversion would have a different impact on the flood hydrograph. Since the main goal of the project was to better understand the effect of wetlands on flood attenuation, the second option was chosen (Juliano and Simonovic 1999). The total amount of diversion can be set equal to the storage capacity of the wetlands.

Once the wetland storage capacity is satisfied, all remaining flow was routed downstream. For modeling diversions, 25% of the sub-watershed peak runoff was set as the diversion rate. Since wetlands covered a maximum of 10% of the watershed area in the scenarios, this assumption was sufficient for filling the volume of the wetlands (Bengtson and Padmanabhan 1999). The flow hydrographs from the earlier landscape simulation runs were used as a guide for developing inflow diversion relationships that are required for diversions, with the timing being adjusted in such a manner that the peak runoff was impacted.

The HMS model was computed to generate runoff for modeled land use and land cover scenarios and curve numbers, future storm events, and various wetland scenarios. Runoff estimates were then used to run the RAS model to evaluate the impacts of urbanization and wetlands on water levels and flood inundation extents.

HEC–HMS has an optimization feature, which was used to calibrate the simulated flow with observed flow. Three parameters were selected for calibration:

1. The curve number

2. Muskingum K

3. Muskingum X parameters

Five different peak events (April 25, 1991; May 25, 1995; May 26, 1996; June 6, 1996; and April 12, 1997) were calibrated using the available data, and the average of those calibrated parameters was taken to develop the final calibrated parameter values (Table 5.5). Initial curve number values were decreased by 10%, the K values were decreased on an average by 33%, and the X values were doubled on average as result of the calibration process (Table 5.5); final parameter values developed were then validated for another five storm events. Table 5.6 shows results for one of those validation events for illustration, and a hydrograph comparison for one of the calibration events is shown in Figure 5.23.

Peak runoff and runoff volume can be matched very closely as shown in Table 5.6 and Figure 5.23, however, it is very difficult to match the time of peak (Figure 5.21) because of the limitation of the SCS unit hydrograph method, which assumes 37.5% of the runoff volume occurs before the peak flow. There are other methods that can be used to transform overland flow to surface runoff, as described in the HEC documentation, but the parameters are difficult to estimate for such a large watershed (HEC 2010c).

Challenges during input data preparation include the resolution of data; for example, more channel cross-sections (i.e., fewer intervals) and more station elevation points are needed, and a higher resolution of DEM is necessary

TABLE 5.5

Hydrologic Modeling System (HMS) Calibration Parameters for Curve Number Scale Factors in All Sub-Basins, with Routing Parameters (K, X) for River Reaches within the Kansas River Basin

	K for Reaches				X for Reaches			
Curve Number Scale	0	1	2	a	0	1	2	a
0.89	13.68	10.09	12.56	8.89	0.40	0.36	0.34	0.34

TABLE 5.6

Peak Runoff, Time of Peak, and Runoff Volume for Hydrologic Modeling System (HMS) Validation Event (i.e., Event on November 19, 1992) within the Kansas River Basin

Event	Peak Runoff (m³/s)			Time of Peak		Runoff Volume (mm)	
	Observed	Simulated	% Diff.	Observed	Simulated	Observed	Simulated
19 Nov 1992	761.6	763.4	−0.24	20 Nov 1992, 17:00	20 Nov 1992, 18:00	4.5	4.4

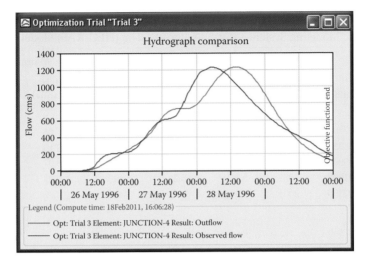

FIGURE 5.23

Hydrograph comparison of observed and simulated flow for calibration event (May 26, 1996) along the Kansas River.

to provide ideal model outputs for the river ecosystem. Note that since this project focuses on evaluating the impact of future land use and land cover changes and the presence of wetlands on peak runoff and the extent of flood inundation, modeling efforts to improve the timing of river crest were subordinated to modeling the relative impacts of land use and land cover change and wetland presence.

Results from HEC–HMS simulations were used in HEC–RAS simulations and additional results for calibration and validation of these models are presented in detail in the next section. The main parameter that was changed to calibrate the model was the roughness coefficient, Manning's n. Manning's n values are generalized to a single value for the channel and the bank areas (the value closest to the banks was selected), as using a Left Bank, Channel, and Right Bank simplification makes the calibration process easier and quicker to conduct. The (five) gauge data were calibrated individually by adjusting their Manning's n values until the water surface elevations matched the reported USGS field measurements (Table 5.7). Then, the calibrated parameters were validated for four different peak events (Table 5.7). Consequently, the calibration results were satisfactory with negligible differences between the actual and the simulated values. These validation results met the needs of the project, with differences in water level being very small (Table 5.7). The overall validation process R^2 and Nash–Sutcliffe coefficient were 0.97 and 0.96, respectively, and therefore the model was used to assess the changes in land use and land cover on flooding for the basin.

TABLE 5.7

Hydrologic Modeling System—River Analysis System (HEC–RAS) Model

Event 2005-06-05	Flow (m³/s)	Water Elevation			Manning's n	
		Observed (m)	Simulated (m)	% Difference	Main Channel	Overbanks
a						
Gauge 06887500	1064	4.55	4.53	0.34	0.06	
Gauge 06888350	1596	5.37	5.37	0.06	0.10	0.14
Gauge 06889000	1913	7.38	7.34	0.58	0.07	0.08
Gauge 06891000	2072	5.18	5.20	−0.35	0.03	
Gauge 06892350	2001	5.99	6.02	−0.41	0.02	
b						
Gauge 06887500						
Event 2001-06-21	880	3.99	4.09	−2.44		
Event 2009-04-27	690	3.63	3.61	0.42		
Event 2008-06-20	908	4.16	4.23	−1.54		
Event 2007-05-07	1049	4.39	4.69	−6.73		
Gauge 06888350						
Event 2001-06-21	934	4.45	4.45	0.00		
Event 2009-04-27	843	4.41	4.29	2.70		
Event 2008-06-20	962	4.63	4.49	2.96		
Event 2007-05-07	1760	5.47	5.57	−1.78		
Gauge 06889000						
Event 2001-06-21	1268	6.02	6.35	−5.52		
Event 2009-04-27	1661	7.00	7.01	−0.13		
Event 2008-06-20	1035	5.51	5.85	−6.08		
Event 2007-05-07	2592	8.75	8.01	8.43		
Gauge 06891000						
Event 2001-06-21	1534	4.46	4.60	−3.01		
Event 2009-04-27	2216	5.36	5.33	0.51		
Event 2008-06-20	1194	3.93	4.11	−4.66		
Event 2007-05-07	3481	6.67	6.24	6.44		
Gauge 06892350						
Event 2001-06-21	2287	6.40	6.33	1.05		
Event 2009-04-27	2157	6.10	6.19	−1.50		
Event 2008-06-20	1106	4.58	4.84	−5.73		
Event 2007-05-07	3396	7.92	7.40	6.62		

Note: (a) Calibration and (b) validation within the Kansas River Basin.

5.2.5.4 Kansas River Streamflow for Modeled Land Use and Land Cover Scenarios

The modeling results for the different land use scenarios (Table 5.4) indicate three future landscape scenarios, as well as the baseline scenario:

1. Scenario 1 (ICLUS output for 2020)
2. Scenario 2 (interchanging low and medium developed intensity classes for 2030)
3. Scenario 3 (interchanging low and high developed intensity classes), shown in Table 5.8

Indications from the landscape models are that runoff values increase with time (Table 5.8) and the 200-mm storm event scenario, as expected, generates more runoff compared to the 175- and 150-mm storm event scenarios. The increase in runoff is small in the 2020 scenario, and increases in 2030 and 2040. The runoff values also demonstrate a marked increase with the shift in time from low intensity development to a higher intensity development of the landscape. The likely reason for the subtle increase between the present and the 2020 scenarios is the small difference between the curve numbers for low intensity development and other land use and land cover classes like cultivated crops, pasture, and forest. Note that the majority of soil types for the modeled areas are the hydrologic groups C or D.

Urbanization potentially caused the increased peak runoff in the models, and increased peak runoff potentially resulted in more soil erosion, which led to increases in sediment loading. Therefore, the watershed is likely more vulnerable to flooding risk and degraded water quality under these urbanized scenarios, suggesting a need to develop more strategies in the landscape to mitigate the adverse impacts of urbanization on flooding.

5.2.5.4.1 Impacts of Wetlands on Kansas River Streamflow

The landscape model was rerun for Scenario 1, Scenario 2, and Scenario 3 after the wetland area was increased to 6%, 8%, and 10% for Scenario 1, Scenario 2, and Scenario 3, respectively. Results for 150-mm storm are shown

TABLE 5.8

Peak Runoff Estimates for Different Land Use and Land Cover Scenarios for the Modeled 150-mm, 175-mm, and 200-mm Storms within the Kansas River Basin

Storm		Peak Runoff (m³/s)			
(in)	(mm)	Baseline	Scenario 1, 2020	Scenario 2, 2030	Scenario 3, 2040
6	152.4	5162.1	5208.7	5551.4	6144.2
7	177.8	6675.9	6723.6	7022.1	7681.2
8	203.2	8252.9	8300.1	8532.9	9250.2

194

Remote Sensing for Landscape Ecology

TABLE 5.9

Comparison of Peak Runoff and Runoff Volume Estimates between Original Landscape Conditions and Modeled Wetland Scenarios for Modeled 150-mm Storm Event in the Kansas River Basin

Scenario	Original Peak Runoff (m³/s)	Peak Runoff with Wetlands (m³/s)	% Diff. (m³/s)
Scenario 1 (6%)	5208.7	4606	11.6
Scenario 2 (8%)	5551.4	4744.2	14.5
Scenario 3 (10%)	6144.2	5060.6	17.6
Scenario	Original Runoff Volume (mm)	Runoff Volume with Wetlands (mm)	% Diff. (mm)
Scenario 1 (6%)	72.7	54.7	24.8
Scenario 2 (8%)	76.9	52.9	31.2
Scenario 3 (10%)	81.5	51.5	36.8

in Table 5.9; as expected, and per landscape ecological theory, wetlands reduce runoff flow and volume. The peak runoff and the runoff volume of the storm events were reduced by varying degrees over the different scenarios (Table 5.9) and a high of 17.6% reduction was obtained for peak runoff while a high of 36.8% reduction was obtained for runoff volume. As the area of wetlands increased, the volume of water that was modeled to be retained within the watershed increased. Peak runoff increased for the future that was modeled to be increased in landscape urbanization, and it was noted that the higher the runoff the higher the flow reduction, as a result of increasing the areal coverage of wetland ecosystems in the landscape. Incorporating wetlands into the modeled future landscape resulted in an average 14% reduction in peak runoff and an average 31% reduction in runoff volume for the Kansas River Basin. It is important to interpret these results with caution, however, in that the actual runoff reduction rate may vary for other watersheds, depending on the unique characteristics of the watershed such as terrain, land use and land cover, and soil type(s) (Huang and Sumner 2011). Other important variability factors include the physical characteristics of the wetlands in the landscape, such as the depth, volume, orientation, vegetational characteristics, and at a broader scale the interspersion/distribution of the wetlands within the larger landscape matrix.

5.2.5.4.2 Kansas River Water Levels for Modeled Land Use and Land Cover Scenarios

The calibrated HEC–RAS model was run for land use and land cover scenarios (Table 5.4) using results from HEC–HMS to obtain water levels and flood inundation extent.

The results at various gauges depict a gradual increase in water levels of the future for the basin (Table 5.10); the increase is small between present day and 2020 scenarios, but then increases in a more pronounced manner for

TABLE 5.10

Kansas River Water Level Estimates for Different Land Use and Land Cover, and Soil Conservation Service (SCS) Storm Scenarios

SCS Storm	Water Elevation (m)			
150-mm	Baseline	Scenario 1	Scenario 2	Scenario 3
6887500	4.68	4.75	4.88	5.03
6888350	5.56	5.62	5.75	5.89
6889000	8.74	8.79	8.9	9.03
6891000	7.01	7.04	7.11	7.18
6892350	8.82	8.85	9.08	9.44
175-mm	Baseline	Scenario 1	Scenario 2	Scenario 3
6887500	5.21	5.27	5.39	5.51
6888350	6.09	6.15	6.26	6.38
6889000	9.36	9.41	9.51	9.63
6891000	7.59	7.62	7.67	7.73
6892350	9.75	9.78	9.94	10.28
200-mm	Baseline	Scenario 1	Scenario 2	Scenario 3
6887500	5.62	5.66	5.77	5.88
6888350	6.57	6.57	6.68	6.81
6889000	9.86	9.91	10.04	10.13
6891000	8.05	8.07	8.12	8.2
6892350	10.54	10.56	10.67	10.98

the 2040 scenario, indicating the increased impacts of urbanization on flood hazard. The highest water levels modeled were obtained from the 200-mm event, which is understandable since it had the highest runoff values. The largest increase in water level was observed for the Gauge 06892350, where the difference between present and 2040 scenarios was 0.62 meters for the 150-mm storm, 0.53 meters for the 175-mm storm, and 0.44 meters for the 200-mm storm (Table 5.10). This decreasing trend illustrates the influence of the topography of the river basin on flood events, typically observed in drainage basins that have a classic funnel shape, where the area of the watershed increases at higher elevations.

5.2.5.4.3 *Impacts of Wetlands on Kansas River Water Levels*

As expected, using calibrated HEC–RAS simulations at various gauges using HEC–HMS, as areal coverage of wetlands increased from 6%, to 8%, to 10% for Scenario 1, Scenario 2, and Scenario 3, respectively, there is a modeled decrease in water levels within the river ecosystem (Table 5.11). The modeled differences in water level is dependent upon the location along the river, and is considerable at all locations, with a maximum decrease of 15.50% and 1.4 meters in the water level being simulated.

TABLE 5.11

Comparison between River Water Level Estimates for Baseline (Original) and Increased Wetland Landscape Scenarios for Modeled 150-mm Storm Event in the Kansas River Basin

SCS 150-mm	Scenario 1, 6% (m)			Scenario 2, 8% (m)			Scenario 3, 10% (m)		
	Original Elevation	Wetland Elevation	% Diff.	Original Elevation	Wetland Elevation	% Diff.	Original Elevation	Wetland Elevation	% Diff.
6887500	4.75	4.71	0.84	4.88	4.84	0.82	5.03	4.99	0.80
6888350	5.62	5.4	3.91	5.75	5.46	5.04	5.89	5.57	5.43
6889000	8.79	7.78	11.49	8.9	7.6	14.61	9.03	7.63	15.50
6891000	7.04	6.24	11.36	7.11	6.22	12.52	7.18	6.14	14.48
6892350	8.85	8.43	4.75	9.08	8.53	6.06	9.44	8.75	7.31

5.2.5.4.4 *Kansas River Inundation Extents for Modeled*
Land Use and Land Cover Scenarios

The landscape area that is inundated within the river basin was modeled in HEC–GeoRAS by using the RAS model results as inputs. The extent of the inundation was calculated and the vulnerable segments of the river basin where flooding is severe was identified in GeoRAS. The inundation area results for the different future land use and land cover scenarios indicate that, with increasing storm depth and urbanization in the future, the flood inundation area within the landscape increases (Table 5.12). An average increase of 58 km^2 is observed between 150-mm and 175-mm storms and an average increase of 49 km^2 is observed between the 175-mm and 200-mm storms, for all three scenarios.

Comparing the inundation area landscape visualization results from different design storms allows for a visual identification of the events, and likely susceptible flooding areas within the basin (Figure 5.24), all based

TABLE 5.12

Inundation Area for Future Land Use and Land Cover, and Storm Scenarios, Developed from HEC–GeoRAS

SCS Storm		Area (km^2)			
(in)	**(mm)**	**Baseline**	**Scenario 1**	**Scenario 2**	**Scenario 3**
6	152.4	350.83	354.82	366.35	379.91
7	177.8	409.16	413.59	424.02	436.27
8	203.2	458.69	461.61	473.11	484.49

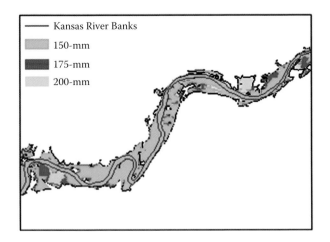

FIGURE 5.24
GeoRAS inundation area landscape visualization results comparing the different storm events: 150 mm, 175 mm, and 200 mm for the 2040 scenario.

TABLE 5.13

Comparison between Inundation Area Estimates Baseline (Original) and Increased Wetland Landscape Scenarios, for Modeled 150-mm Storm Event in the Kansas River Basin

SCS 150-mm	Original Area (km²)	Area with Wetland (km²)	% Diff.
Scenario 1 (6%)	354.82	300.13	15.41
Scenario 2 (8%)	366.35	301.33	17.75
Scenario 3 (10%)	379.91	306.90	19.22

upon the simulated flooding along the Kansas River for the different storm scenarios in 2040. The locations at which the 175-mm and 200-mm scenarios begin to have a significant impact on human safety and well-being is easily observed in Figure 5.24. Such information is what was desired by the clients of this project, who found it most useful to utilize the GIS data product to make initial decisions about land use planning and conservation practices, even if simply confirming the concepts that they have been putting into practice for several years (e.g., wetland conservation in riparian areas). These outputs also potentially catalyze public discussions about the opportunities to reduce flooding risk by incorporating wetlands back into altered landscapes of the Kansas River Basin, or elsewhere among the 22 tributary watersheds of the Missouri River and Mississippi River in this project. Potentially flooded areas could also be planned for, and avoidance strategies could be developed, for sustainable planning of agricultural and residential developments, as well. As one might expect, based upon the prior results, wetlands can be used to reduce flood inundation (Table 5.13). Comparison of inundation area between baseline and increased wetland landscape scenarios for the 150-mm storm event indicates that the inundation area is reduced on average by about 17.5% in the study area.

5.2.5.5 *Importance of Remote Sensing for Landscape Ecological Scenario Development*

The Kansas River Basin landscape models, using HEC–HMS and HEC–RAS, exemplify how a thoughtful and creative synergism can occur among remote sensing scientists and practitioners, hydrologists, and landscape ecologists, all by utilizing geospatial models within a landscape to determine the effects of future landscape condition on communities. Large watersheds and landscapes are complex, however skill and creativity in the use of existing data and models can be brought to bear on important landscape ecological issues (Conagua 2015), and in this project such efforts had a positive impact on the public's and decision makers' understanding of the watershed in which they reside, and potential for conservation planning in the future.

Modeling work utilizing the ICLUS projections of urban land use densities for 2020 was integral to understanding how the impact of human activities will determine landscape condition over the coming century. The design storms with a 100-year frequency for the region allowed watershed modeling to occur that effectively evaluated the critical physical importance of wetlands in the landscape, as well as their influence on flood mitigation/attenuation in the future of the Great Rivers of the United States.

Specifically, the results demonstrate an impressive potential increase in peak runoff and flood inundation extents for the various future (predicted) land use and land cover scenarios, with about a 1,000 cm increase in peak runoff for the different storms from the present to the 2040 scenario as a consequence of increasing urbanization extent in the landscape and increase in density of that urbanization. There is an approximate 10–19% increase in peak runoff and a 2–7% increase in water level for all the land use and land cover, and design storm scenarios within the basin, given the future scenario of urbanization increases, in addition to flood inundation extents for the watershed. The increased predicted inundation area of 5–8% between the present and 2040, given the current trajectory of urbanization suggests that greater adverse impacts on both agricultural and urban areas along the river are likely, increasing the hardships on the human population of the region that have occurred in the past, including continued economic losses in the area. Accordingly, flood-vulnerable areas should be avoided for agricultural production and residential development, and wetland mitigation is recommended, per the model results, for the river basin.

Because wetlands were capable of causing a sizeable reduction in peak predicted runoff and inundation extent, an increase in their presence in general would likely result in a reduction of flood volumes by 25–37%, per the model results. Modeling the different wetland acreage scenarios resulted in an average decrease of 8.6% in the water levels as well, and with a maximum decrease of 15.5% at one particular location and a decrease of 15–19% in the area of inundation, it is apparent that the presence of wetlands would reduce the risks of flooding to urban, suburban, and extra-urban residents, as well as associated structures.

The landscape models in this project can be used to test the impacts of land use and land cover changes, rainfall predictions, and channel modifications in other watersheds, with some additional calibrations. However, as always with landscape scale analyses, the models are designed for relatively broader scale decisions, and if the results are applied to a small segment of a watershed they may not retain model accuracy. It should be noted that these models are conservative in terms of risk, in that there is also a possibility that the 100-year 24-hour storm values used in this study might be different in the future as a result of ongoing climate change, which could result in higher surface runoff values. Additionally, wetlands were assumed to be clumped together, rather than in some more widely distributed manner throughout the landscape, within the HMS modeling scenarios. This assumption affects the peak runoff, which may actually be higher, if wetlands were more widely distributed in the landscape.

Because results from this project are site-specific, actual increases or decreases in peak runoff and runoff volume due to various land use and land cover changes, such as changes in wetland configurations in the landscape, may vary for the other 22 tributary watersheds (Figure 5.7), or elsewhere, depending on the unique characteristics of the watershed and the wetlands constructed. Although beyond the scope of this project, the results of this work could be paired with an economic analysis to determine whether the savings in damages obtained from flood reductions, as a result of modifying and increasing wetland volumes, justify that aspect of the costs of constructing and maintaining those wetlands, as well as take into account the other ecosystem services provided by those wetlands. Environmental managers and practitioners in USEPA Region 7 have evaluated this approach to prioritize and inform decision makers about the ecological and societal benefits of ecosystem restoration in specific watersheds, and for the implementation and assessment of Best Management Practices within the current land use areas of the Kansas River floodplain and other river ecosystems of the Midwestern United States (Lopez, in preparation).

In this penultimate case study, which builds upon the prior-discussed landscape ecological water quality modeling work in the Ozark Mountains, we demonstrate how remote sensing, field-based metrics, and geospatial modeling can be combined to provide the ideal synergy of technology and science to meet the needs of society and communities, allowing for more effective watershed planning choices and natural resources management decisions. It should be mentioned that, in the previous chapter, we introduced two other excellent examples of similar approaches, which certainly differed with respect to the relative abundance of available field-based data throughout the immense project area (i.e., the Great Lakes Basin), yet enabled the fundamental analysis of broad-scale assessment of conditions, through the availability of remote sensing–based data for the entirety of that basin. The bicoastal example of future landscapes in California and North Carolina coastal regions also required creative modeling tools to allow for effective predictions of the potential impacts of sea level rise in the future.

Each of the case studies, taken by themselves, provides excellent lessons and details on the effective use of remote sensing for landscape ecology, exercising the essential principles laid out earlier in this book, and intending to enable the reader to imagine a future that is bright and plentiful with opportunities to apply similar techniques to additional challenging questions. As time proceeds, we continue to see innovative technologies and methodologies that may apply to these future questions too, and indeed several key challenges remain for us all to work toward overcoming. In the final chapter of this book, we will explore several of these challenging areas that remain, and look at these challenges with our eyes open and minds prepared to tackle them. The future of remote sensing and landscape sciences is abundant with exciting new avenues for improving the lives of people by utilizing the technology and informational progress gained thus far within these disciplines.

6

Meeting the Landscape Ecology Challenges of the Future with Remote Sensing

The main goal of this book has been to provide a full spectrum of landscape ecology approaches, methods, and metrics that are possible for today's available remote sensing data, specifically as it applies to information needs for environmental practitioners and decision makers, and taking into account spatial and spectral resolution, as well as practicality for the applications considered and the landscape conditions we observe today. The several additional goals of this book were to provide examples that were comprehensive in that they provided a number of opportunities for informing analogous work, and likely much of that work could apply to many other geographies around the world, utilizing the diverse variety of remote sensing platforms and data available now, and increasingly in the coming years. The intention was to provide a "ramping up" of complexity, from the Laurentian Great Lakes and coastal North Carolina's and California's expansive areas and drivers of change, to the more detailed calibration of landscape models within the Ozark Mountains and Great Rivers Watershed. Notwithstanding the variable geographic extents and modeling complexity of the case studies, the goals remained the same among the various projects, that is, utilizing a combination of technology, practical applicability, and theoretical considerations for improved analysis and conveyance of the ecology of large landscapes, utilizing real-world data for various gradients of change. Several expansive areas of active and future inquiry are included in this chapter to complement the case studies of Chapters 4 and 5, based upon societal directions and needs that are emerging presently and making themselves apparent in the popularity and success of landscape collaboratives for restoration, water quality programs and strategic watershed management groups, and sea level rise monitoring and adaptation planners and practitioners; all benefited by the indispensable use of remote sensing technology and science. The following areas of emerging inquiry and information needs are consequential in that they all may result in new disciplinary areas upcoming for the landscape sciences and remote sensing sciences, as well as the multitude of newly trained professionals of the future.

6.1 Future Trends in Landscape Sciences for Inventory, Monitoring, and Assessment

Landscape sciences and, more specifically, landscape ecology has undergone tremendous change in the past two decades, and the melding of these relatively young but maturing disciplines has yielded new and exciting trends in inventory, monitoring, and assessment. New issues, new results, and advances in techniques, as well as innovative new applications for the landscape sciences and landscape ecology, all have a positive future because of their increased applicability to the fundamental needs of society, especially in terms of assessment, planning, and decision making. Simultaneously with this increased need for compelling information about ecosystems and their status or conditions, the coming decade will likely witness important new developments and refinements of earlier paradigms, enhancing our understanding of larger areas of the Earth and obviating the need for managing land as a landscape, regardless of other boundaries, and developing new ways of combining these (much more complex) analyses and articulating them to the public in (likely nearly) real-time, ensuring that the lives and the needs of diverse communities are incorporated into outputs from project work (Figure 6.1).

FIGURE 6.1
The future of remote sensing and landscape ecology is rich with opportunities to serve the needs of diverse communities around the globe, and outputs should be geared toward meeting these needs in order to remain relevant to decision makers. (Image: Subsistence fishing villager in Kompong Phluk, Cambodia.)

Of most importance in this new landscape approach to decision making for the future is the desire and ability to incorporate all processes, both ecological and sociological, for improved analyses and decision making at multiple scales and time frames. The landscape sciences currently represent the culmination of a number of decades of intensive study and theoretical development, and the application of practical geospatial technologies to much of what society now relies upon or is striving for, worldwide, such as safe drinking water, sufficient open spaces for the public, and a sustainable landscape that can be relied upon for generations. The ultimate result is a set of concepts that can be used now, and into the future, to formulate innovative new approaches for solving the environmental challenges of the future, such as overpopulation, loss of green spaces, and the impact of environmental conditions on food security (Lal and Stewart 2012). Despite such challenges, the use of in situ data and remote sensing data from airborne and satellite platforms, in combination with mathematical or geostatistical approaches, can be used to ensure a conceivable path forward for people's future, as we have demonstrated in a number of case studies in this book. However, to successfully predict the future trends and the risks that are posed by natural and anthropogenic drivers of change, we must expand even further our scope and reach of work.

An approach that is currently used in the landscape sciences, and may be increasingly relied upon in the coming years, is that of environmental risk assessment. Risk assessments involve the same methods utilized in Chapters 4 and 5, with regard to sea level rise, water quality, and watershed dynamics, by the characterization of specific risks and how those risks may influence or expose natural resources, such as any ecosystem type, or receptors of the risk, such as the organisms within a forest or people living within a landscape. The exposure of an ecosystem or organism (the receptor) to a natural- or human-induced stressor may often define the problem that is necessary to address through decisions and actions. Risk can be determined by the characterization of exposures and the ecological effects to receptors by stressors. By developing our future capabilities of determining the condition of both resources and receptors, efficiency and effectiveness, landscape ecology projects will increase, particularly as sensors and models advance to identify more and more risks and their likelihood of occurrence (e.g., Asner et al. 2016; Sarkar et al. 2016; Wang et al. 2016).

6.2 Emphases on Watershed Restoration and Coastal Planning

Among the many possible risks to be called upon by society to assess and manage, more and more of these risks involve the issue of water quantity. Flooding is a major natural hazard that impacts different regions across the world every year. Between 2000 and 2008, among the various types of natural hazards, floods have affected the largest number of people worldwide, averaging 99 million people per year (WDR 2010). Within the United States from 1972 to 2006, the property losses due to a catastrophic flood event (a flood causing damage of $1 million or more) averaged approximately $80 million (Changnon 2008); on average, floods kill approximately 140 people each year in the United States (USGS 2006).

As alluded to in Chapter 5, over the past two decades in the United States, the Missouri River and Mississippi River Basins have experienced unprecedented flooding that caused fatalities, evacuations, and large financial losses. Climate change is expected to continue the enhancement of the risks of extreme storm events (Milly et al. 2002) and increases in the frequency of flash floods. Large-area floods in many regions of the world are very likely to continue to occur in the future (Alley et al. 2003; Parry et al. 2007).

In addition, the world is also undergoing the largest rate of urbanization in history. In 2008, for the first time in history, half of the world's population resided in urban areas, and urban areas are expected to experience most of the future global population growth. From 1982 to 1997 the amount of land devoted to urban and built-up land uses in the United States increased by 34% (USDA 2001). As was noted in earlier chapters, urbanization can increase the size and frequency of floods along river courses, and may also expose those communities to increasing flood hazards (USGS 2003), thereby increasing the future focus by planners and land managers on the role of urbanization in the prediction of flood levels and damage, primarily for disaster management and urban and regional planning (Milly et al. 2008).

As analyzed in detail in this book, a key example in the Midwestern United States, the Kansas River, has been prone to flooding over the last century. It has faced two significant flood events in 1951 and 1993. The damages from the 1951 flood were substantial. Nineteen people were killed and about 1,100 injured with total damage estimates as high as $2.5 billion (Juracek et al. 2001). During the height of the flood, on July 13, 1951, nearly 90% of the flow in the Missouri River at Kansas City came from the Kansas River, a tributary comprising only 12% of the Missouri's drainage basin (Juracek et al. 2001). The historic 1993 Midwestern floods affected nine U.S. states and resulted in $15 billion worth of flood damages. Such flooding events are likely to continue and increase in size, intensity, and stochasticity into the foreseeable future, worldwide, as a result of climate change.

An effective approach for assessing flood risks for people and their property within a watershed, such as along the Kansas River, is through the production of flood risk models, such as those in this book, which show areas prone to flooding events of known return periods. To mitigate flood risk, reservoirs and levees have been historically utilized, and this trend is likely to increase more in the coming years as ponds' and wetlands' role in flood attenuation and water quality improvement increases worldwide (Mitsch et al. 2001; Acreman and Holden 2013; Ludwig et al. 2016). Wetlands also have the capability of short-term surface water storage, and can reduce downstream flood peaks, thus serving people very far downstream.

In coastal areas around the world, millions of people live within kilometers of the coast, and make up a large percentage of the world's population. Therefore, extreme change is a critical concern for coastal marine planners and other decision makers in coastal areas of the world, including developed areas that are intimately linked to surrounding tidal and estuarine wetland areas. The extent and the degree of the loss of marine coastal wetlands is an important and challenging question for geographers, ecologists, and policy makers alike, which will need to be answered, particularly in coastal areas where the ecological and societal impacts of changing sea levels are fairly immediately threatening infrastructure, and adaptation strategies are needed in the near-term; the addition/restoration of wetlands in these coastal areas is one of the likely solutions. Chapter 4 outlined the critical nature of such marine coastal wetlands, and the need for technologies and scientific theory to address these issues will become a very pressing issue as the twenty-first century proceeds, and sea levels continue to rise, as is predicted by IPCC models. From a tidal marsh management perspective, coastal wetlands along marine coasts are among the most susceptible of ecosystems to changes in sea level rise, and associated changes in sea water surge intensity, and surge heights that result from hurricanes, typhoons, and tsunamis. Some of the impacts on marine coastal wetlands, and upland areas that abut them (where people often congregate), from sea level rise and surge are a result of wetland's particular sensitivity to short-term and long-term changes in inundation, salinity, and soil characteristics. Therefore, wetlands will likely continue to be a critical ecosystem of interest in the fields of remote sensing and landscape ecology for the coming century, for the reasons outlined throughout this book.

As outlined in detail in Chapter 4, there are a number of landscape approaches increasingly being used to measure and monitor the effects of SLR. Noting the importance of the proper and relevant scale of a management problem, and resolution of information necessary to have an impact on associated decision making, are the main drivers of a landscape scale ecological project's cost. Currently, and likely to become even more important in the future, the drivers of project costs will become even more important in order for a project to be funded, and successful project proposals now and into the future will continue to require assurances of quality data and outcomes, particularly as it pertains to ecological models, as well as cost-effectiveness of the overall products and outcomes.

These are just a few key applications that demonstrate a wide breadth of applications where landscape metrics can potentially provide solutions for society and those natural resources that underpin a growing, and urbanizing, global society into the future. It is recommended that landscape ecologists, remote sensing scientists, geographic information specialists, and personnel from governmental agencies who are engaged in landscape science research to continue to utilize the wide variety of landscape metrics available, and to also push the boundaries of developing new metrics and models as needed and conceived, especially where they are newly applicable and useful for better understanding the multitude of highly complex ecological processes. As urged in the first edition of this book, the wide array of scientists involved in the endeavors of landscape ecology who utilize remote sensing must continue to learn from the principles and applications that each field provides (Frohn 1997). As in the first edition, we continue to be reminded that, perhaps, the most important lesson to be learned is that landscape ecology, remote sensing, and geographic information sciences each have extremely important and unique characteristics that are each very necessary, and each discipline can thus contribute to the development of better ways to model, monitor, and assess the ecosystems of the world into the future, synergistically.

6.3 The Importance of Ecological Goods and Services for Communities

Currently, national and regional surveys of ecosystems assess the distribution and ecological conditions of ecosystems and, much less commonly, the ecological functions of ecosystems. In order to better understand these ecological functions, and their importance to a functioning and healthy society, the importance of *ecosystem services* (used interchangeably with "ecological goods and services") must be better understood within the communities that benefit from these natural areas and resources.

A better understanding of the connection between what has always been described as ecological functions, with the production of ecological goods and services, allows us to better meet basic human needs into the future, and determining this connection is a major challenge currently facing environmental and resource management professionals. The landscape sciences are well suited to assist in these goals, and the use of remote sensing will continue to enable the mapping and modeling of these ecosystem services in the coming years, likely, more and more. A major effort to map ecosystem services nationwide in the United States has produced some promising results, and is discussed briefly later in this chapter. Relatedly, and in a globally reaching and seminal effort to synthesize the extent and scope of all ecosystem services, the Millennium Ecosystem Assessment (MEA 2005) has provided tremendous leadership in summating ecosystem services, including a series of reports and descriptions of ecosystem services that span a wide variety of themes, such as a detailed synthesis of water and wetlands, which has also brought wetland ecosystem services to the forefront of global environmental resource management interests. This approach is continuing to produce an emerging and clearer articulation of the many fundamental ecosystem services that abound, among all ecosystems on Earth, as outlined in the following section.

6.3.1 Ecosystem Supporting Services

6.3.1.1 Carbon Cycling

Ecosystems are significant carbon reservoirs and contribute to regulating global climate change through sequestration and release of fixed carbon. Carbon is contained in the standing crops of vegetation and in litter, organic soil/sediments. The magnitude of storage depends upon ecosystem type and size, vegetation, the depth/type of soils, groundwater levels, nutrient levels, pH, and other biogeochemical factors (Reddy and DeLaune 2008), necessitating analyses of such factors. These carbon reservoirs may supply large amounts of carbon to the atmosphere if water levels are lowered or land management practices result in oxidation of soils (Figure 6.2). Because ecosystems serve as significant carbon sinks, the destruction of ecosystems will release carbon dioxide. Although much is still not understood about the role of ecosystems in the global carbon cycle, it is generally agreed that the gain/loss of ecosystems is a major input to global carbon budgets, thus

FIGURE 6.2
Ecosystems are significant carbon reservoirs and contribute to regulating global climate change through sequestration and release of carbon, one of the several critical ecosystem services that contribute to human well-being. (Image: Tropical forest, Central Laos.)

affecting the amounts of carbon dioxide and other greenhouse gases in the atmosphere, with potential for contributing to climate change.

6.3.1.2 Wildlife Habitat

Ecosystems are a reservoir for biodiversity, which has many links to human well-being. Estuarine and marine fish and shellfish (see Provisioning Ecosystem Services, below), various birds, and certain mammals require ecosystems to survive. Many of the U.S. breeding bird populations (including ducks, geese, woodpeckers, hawks, wading birds, and many songbirds) feed, nest, reproduce, and raise their young in ecosystems. Migratory waterfowl use certain ecosystems as resting, feeding, breeding, or nesting grounds for at least part of the year.

6.3.2 Ecosystem Regulating Services

Ecosystems protect human well-being by mitigating floods and buffering the effects of coastal storms. Ecosystems reduce flooding by absorbing rainwater and by slowing the downstream flow of floodwater. Ecosystems also function as natural attenuators of water flow, slowly releasing surface- and groundwater. Trees, root mats, and other vegetation also slow the speed of floodwaters and distribute them more slowly over a floodplain, lowering flood heights, and affecting erosion/accretion in the surrounding landscape. Ecosystems decrease the area of open water (fetch) for wind to form waves, which increases drag on water motion, thereby decreasing the amplitude of waves or storm surges. The value of ecosystems to reduce impacts of floods and storms has often been retrospective, based on the estimated costs of damage or loss after a flood or storm has occurred.

6.3.3 Ecosystem Provisioning Services

Many ecosystems contribute to recharging groundwater aquifers that are an important source of drinking water and irrigation. Plants, microbes, and soils in ecosystems affect water quality by removing excess nutrients, sediments, and toxic chemicals, therefore improving water quality for humans and aquatic biota. Ecosystems intercept surface water runoff from higher dry land before the runoff reaches open water, reducing eutrophication in downstream waters and preventing contaminants from reaching groundwater and other sources of drinking water.

Many fishing and shell-fishing industries harvest ecosystem-dependent species. Most commercial and game fish breed and raise their young in coastal marshes and estuaries. Menhaden, flounder, sea trout, spot, croaker, and striped bass are among the many familiar fish that depend on ecosystems. Shrimp, oysters, clams, and blue and Dungeness crabs likewise need these ecosystems for food, shelter, and breeding grounds.

6.3.4 Ecosystem Cultural Services

Ecosystems also have recreational, historical, educational, aesthetic, and other cultural values that are held by the public and vary by geographic area and the societal values of the particular population of humans in an area.

The importance of all of these ecological goods and services are apparent in almost every venue of public discourse that occurs where human planning and decision making is occurring, in a corporate environment, governmental agency, or grassroots community meeting. Landscape ecologists have an ongoing, and likely increasing, role in these discussions, and indeed often a leading role in explaining them from a scientific perspective. In the coming years, collaborative work among a number of other disciplines (whether they are rural sociologists, hydrologists, image analysts, archeologists, ethnobotanists, or one of the many other disciplines of import) will enable an expanded conversation about how these ecological goods and services can be mutually beneficial and engaged by all participants.

6.4 Using Remote Sensing to Map Ecosystem Services

Ecosystem services may be linked with the critical environmental science needs of decision makers and communities, which improves the well-being of all members of the public and also assists in demonstrating the value that environmental science and natural resources management provides to the public. One example of how this connection between remote sensing data, landscape ecology, and ecosystem services is the U.S. Environmental Protection Agency's National Atlas of Ecosystem Services (NAES). USEPA has continued to provide definitive value to stakeholders across the country by leading on this project, which provide a user-friendly Internet decision support tool that takes the atlas to a new level of functionality and science communication. The following URL directs the reader to this tool, which provides remote sensing–based data in a format that assists in the envisioning of ecosystem services for society to better understand, and allows the catalyzing of additional thinking about the topic of ecosystem services, from a landscape perspective: https://www.epa.gov/enviroatlas/ecosystem -services-enviroatlas (checked January 25, 2017).

One notices, upon exploration of the National Atlas of Ecosystem Services, that it has great capabilities and even more potential, and perhaps analogues for other organizations, for expansion and refinement with new remote sensing data and secondary products, at a variety of scales. In general, the NAES provides an exciting new avenue for the utilization of remote sensing sciences in the context of landscape ecology, since those data (such as are described in Chapters 4 and 5) are foundational to the approach taken in the NAES.

The development and implementation of nationwide maps (i.e., geospatial/ statistical models) of ecosystem goods and services, focusing on water provisioning, air quality, nitrogen retention, water quality, biological diversity, carbon storage/sequestration, food/fiber, and urban ecology is realized with this developing tool, which can be used in a number of applications for science, natural resources management, planning, education/outreach, and likely many other applications. The NAES project was conceptualized and implemented to directly address the interdisciplinary exposure research missions of USEPA and to address the goals of EPA's Office of Research and Development's Ecosystems Research Program, utilizing the foundational work of the Millennium Ecosystem Assessment (2005), with a strong emphasis on collaboration with many other national and international partners. The final products of the NAES project are continuously published in the peer-reviewed literature and hosted/updated online by USEPA (https:// www.epa.gov/eco-research/ecosystem-services, checked January 25, 2017) and by collaborators at the USGS and the National Geographic Society.

It is important to note that the original concept for the NAES came out of a general need for better understanding of the linkages between human well-being and ecological conditions across large landscapes, and leverages the metric browser format, articulated earlier in this book, for example, Ozark Mountains, Great Lakes Ecosystem, and Great Rivers (Missouri River and Mississippi River) Watershed case studies. Accordingly, the NAES uses a similar framework used throughout this book, and elsewhere, that relies on a framework of ecosystem "condition" and "drivers of change." The NAES, importantly, builds upon this backbone of ecosystem "condition" and "drivers of change," in a new way, focusing on the supply and demand functions of ecosystem services, thereby providing a linkage between ecosystems and the well-being of humans in the landscape. In this way, the work of the NAES, as specifically articulated in the Millennium Ecosystem Assessment (2005), completes the connections between remote sensing data and technology, landscape ecological theory, and human well-being.

Perhaps most importantly, and looking to the future of the alliance between remote sensing science and the discipline of landscape ecology, is the promise and utility of the NAES for catalyzing conversations and planning that this type of tool can generate. Some of the tools that may come out of the work of the NAES are likely to be optimally inclusive of traditional science-oriented people/organizations, including those people and organizations that may be focused on business, engineering, traditional ecological knowledge, and the many other diverse perspectives that exist. Luckily, the NAES concept, which depends heavily on web-based tools, was prototyped in 2008 intentionally to ensure easily accessible maps that contain important information for communities and people, ideally so that they could be customized and explored from the user's perspective and this format can be adapted to meet other needs, depending upon the community focus and decision maker's goals. Such a community-based engagement approach is an integral component of the evolution in this type of ecosystem services work, which utilized geospatial data sets and the input of communities to explore particular societal aspects of the environmental sciences and natural resources management. The use of remote sensing is an integral component of this type of tool, for the purpose of better understanding ecological processes across large landscapes in the future.

6.5 Moving toward a World of Sustainable Landscapes

The pervasive use of the term *sustainability* has, similar to *ecosystem services*, swept the globe, particularly in the past decade. The concepts embodied within the concepts of sustainability are not new, however, their current incarnation unifies social and ecological perspectives of nature and makes possible a new and positive perspective so that societies have the capability to meet the needs of the present generation while living within the carrying capacity of existing and supportive ecosystems, all the while providing future generations with opportunities to meet their own needs. This viewpoint and concept has had a tremendous effect on the need for quantifying ecosystems and their elements (i.e., identification and characterization), their condition (i.e., ecological functions), and their role in society (i.e., ecosystem services) (Reid et al. 2010).

From the most general concept of sustainability comes a broad view of environmental and ecosystem management issues, which may offer a way to go beyond technological solutions to environmental problems by integrating social participation and policy dialogue with ecological inventorying, monitoring, and assessment activities. These concepts, codified by many of the world's representatives to the 1992 United Nations Conference on Environment and Development, provide us all with an impression of the tremendous magnitude, promise, and challenges that sustainability can bring to the management of the world's natural resources.

The global emphasis on sustainability brings with it a strong message of harmony with nature and with each other. The concepts embodied within sustainability also bring with them a number of other aspects of the appreciation of nature, depending upon the cultural or other values of a community, including a variety of concepts. These concepts can be associated with personal or group value systems, which consider the immutable rights of animals and plants, as well as the sacredness of physical sites throughout the natural world. The messages that are part and parcel of sustainability, that is the universally appealing idea of being caretakers of nature and living in harmony with nature, have potential for encouraging and facilitating the efficient use of resources, and also offer up a tremendous diversity of opinions due to the ever-increasing speed and intensity of cultural homogenization.

Nevertheless, although to a matter of degree, depending upon particular cultural values and societal norms, the differential recognition of the pure rights of the environment has universal appeal and, thus, global significance. Such approaches to environmental and ecosystem management would require preserving the local environmental values held by individuals, including all or a combination of several spiritual, cultural, and personal beliefs, while maintaining a simultaneous view of the global, and cumulative/collective, impacts of environmental decisions and actions.

As alluded to above, the current era, which has now been deemed by some as the Homogenocene, as Mann (1997) defines it: "a new biological era … mixing unlike substances to create a uniform blend," provides us all with numerous challenges for recognizing these different value systems when making environmental management decisions. Considering the effects of this melding impact over the past 500 years or so, it stands to reason that a more mindful melding of the multitude of global value systems, and preserving and respecting diverse cultural and religious value systems that exist throughout the globe, could lead humans on a path toward greater harmony with nature, and perhaps each other.

In a world that is trending toward homogenization of its people, cultures, languages, and ecological systems and functions, this global path to sustainability, which includes cultural values, may be extremely difficult without intentional efforts. Agreements are necessary to engage in the intercultural or interfaith dialogues necessary to approach the precepts of "sustainable use" or "the alternatives to use" to be successful resource managers, planners, and decision makers in the future. There are numerous examples of global environmental issues that are truly impending crises, and hence they must be dealt with thoughtfully and effectively if societies are going to live in true harmony with nature, which is a fundamental component of sustainability. A delicate balance of globalism and inclusion of critically necessary local and regional perspectives may be required to develop robust, adaptive, and respectful formulations of natural resource management plans. Naturally, ecosystems and the landscapes they are embedded within are all very important elements of that planning.

Developing countries face the challenge of reducing poverty and attaining sustainable development at the same time. They have a particular need (and incentive) to benefit from this unifying concept on a global scale. A key link in this global concept and perspective is the wise use of geospatial information, particularly for natural resource characterization and other assessments, including ecosystems, due to their ubiquity across vast landscapes and their importance in delivering crucial ecosystem services.

The emerging critical and future threats to ecosystems and related natural resources, and the sustainability of ecosystems, are most notably soil loss and degradation, water scarcity, and the loss of biological diversity (Running et al. 2004), regardless of the sociological contexts. The perceptions of these environmental problems vary tremendously, depending upon a number of socioecological factors. If geospatial information is to have an impact on the users in these areas, the information they produce needs to be compelling, accurate, and easily accessible to the user (i.e., must have high impact, availability, and perceived relevance to communities and decision makers). Some argue for an approach that addresses this complexity as a multilevel stakeholder approach to sustainable land management, for finding feasible, acceptable, viable, and ecologically sound solutions at local scales. A number of international programs and bilateral cooperation projects have also taken this perspective and started using these sustainable land management (SLM) approaches, either explicitly as in the case of United Nations Capital Development Fund or at least implicitly in their other programs. The SLM approach uses management as an activity on the ground, using appropriate technologies in the respective land use systems. This SLM sustainability paradigm requires that a technology follow five major sustainability principles:

1. Ecological protection
2. Social acceptance
3. Economic productivity
4. Economic viability
5. Risk reduction

Accordingly, a technological approach to resource management that is sustainable would have to be developed using criteria for a particular and locally relevant land use, and would likely not be similarly applicable everywhere. This SLM method follows on an earlier discussion about transdisciplinary approaches, in that it encourages the full exploration and consideration of all dimensions, particularly the economic dimension, but also the social, institutional, political, and ecological dimensions of the community in question. Interestingly, for resource managers, this new local-to-global viewpoint of sustainability creates an entirely new dimension, which has been generally referred to as the realm of the stakeholder.

A number of global environmental professionals have also suggested there is an efficacy to explicitly linking research on global environmental change with sustainable development (Reid et al. 2010), which would necessitate the increased use of remote sensing and geospatial analyses for monitoring ecosystem conditions across vast landscapes, and also for measuring feedback loops between environmental conditions and societal values and activities. Remote sensing data could thus be incorporated into large and complex models of socioeconomic systems to help determine the social and economic impacts of ecosystems and other environmental components or changes. The data could also be integrated with existing efforts, such as GEOSS (Office of Science and Technology Policy 2010), described in the next section, so that such products would best meet the needs of decision makers about sustainable development issues and programs of the future, around the

FIGURE 6.3
The concept of sustainability has taken many forms, and follows the precepts of many island cultures around the world, which are emblematic of the emerging model of sustainability, where culture is an overarching influencing force on the societal, environmental, and economic components that drive environmental science and natural resource management decision making. (Image: The Island of Kahoʻolawe, Hawaiʻi, United States.)

world. An emerging twenty-first century paradigm of sustainability is one that goes well beyond the concept of "intersection" among the environment, society, and economy, and even further than the more interdependent model of sustainability which recognizes a direct dependence of the economy and society upon the environment. This new paradigm of sustainability is one that recognizes the interdependence of the environment, society, and economy, and then overlays the paramount control function of all of those interdependences (i.e., among environment, society, and economy) of "culture," which encompasses them all (Chirico and Farley 2015). This new paradigm of culture-based interdependence of the environment, society, and economy is much like the ancient Hawaiians, and other island-based societies around the globe, who have a deeply engrained value system that is based upon the paramount importance of sustainable utilization of resources, respect for those resources, and rules that guide their societies toward such sustainable (and practical) practices (Figure 6.3).

6.6 Global Perspectives for Systems Analyses of the Future

The diversity of technology and infrastructure to monitor environmental systems, which are inclusive of all landscapes and the totality of the environment (including both the global environmental systems and the social systems, i.e., the totality of global systems) in which they reside, from global to local scales, is growing rapidly as public and private organizations increase their investment in them. Resource planning, decision making, and management requires environmental data that specifically answers the operational details of emerging and unforeseen real-world problems, some of which we have discussed in detail in this book. Much of the techniques available now for routinely addressing some of these topics are much more understood by the public than in the past, partly due to developments in visual media for conveying these data (e.g., maps, models, processed airborne or satellite remote sensing data, and online decision support tools). Therefore, good environmental planning and management requires continued excellent data and outputs to ensure compliance with expectations of the public and to monitor for project or program successes, or emerging difficulties. This has become the expectation, rather than the exception, when developing and producing landscape ecological outputs in large landscapes. Remote sensing products (including aerial photography, along with airborne and satellite imagery) are so routinely used for monitoring and reporting these requirements now, often established by national policies and international treaties, conventions, and agreements (Backaus and Beule 2005), that they are indeed considered common tools shared worldwide. With the advent of GoogleMaps, BingMaps, GoogleEarth, and a host of other easily accessible and user-friendly geospatial data viewers, apps, and online websites, the public at large now shifts their attention from the unique nature of having these data available to them to the expectation of these data being of high caliber and available to them for day-to-day decisions, often in real time.

Remote sensing is frequently used at local to regional scales to improve land management policies and associated decision making processes, especially in areas that are highly agricultural and that are urbanized (Miller and Small 2003; Bryan et al. 2009). For example, data from AVHRR (Advanced Very High Resolution Radiometer), MODIS (Moderate Resolution Imaging Spectroradiometer), and MISR (Multiangle Imaging Spectroradiometer) satellite sensors have been used to regularly monitor biomass burning in Brazil (Koren et al. 2007). Other examples include satellite-based sensors that can be used to monitor and assess important and incremental ambient or diffuse globally significant environmental conditions, such as those conditions related to air pollution (Engel-Cox and Hoff 2005) and urbanization, and the risks associated with those conditions (i.e., human health, or the potential for impacts to people and built structures from flooding or fires) (Miller and Small 2003). Other satellite platforms are proving to be especially useful for the management of urban areas with regard to environmental risks around the globe, such as the relatively fine resolution IKONOS and Quickbird satellites, as well as NASA's Shuttle RADAR Topographic Mapping (SRTM) mission (Miller and Small 2003). These advances are examples of technologies that have a global impact on society, from the perspectives of health and well-being, to the economy of smaller communities and regions. Similarly, the more traditional approach of using historical aerial photography and satellite imagery allows us to establish a historic baseline of local and regional land cover, which can be customized for the needs of decision makers and other audiences, and such historical information can provide the important context needed to guide management policies of the future. The concept of "alternative futures" scenario building, designed by local decision makers and community members (Baker et al. 2004; De Leeuw et al. 2010), is based upon visioning a variety of future scenarios, and knowing the past condition of a landscape helps participants in future exercises. Developing such alternative futures can help many small communities adapt to global environmental and other changes, however, the costs of conducting such futuring exercises is variable, depending upon the availability of remote sensing data and technical capabilities for utilizing the data to develop future scenario models.

At the globally relevant scale, impactful issues of the future must also be effectively addressed into the foreseeable future, and the environmental dynamics of those focus areas need to be understood by appropriately scaled remote sensing data (Running et al. 2004; Reid et al. 2010). For example, satellite imagery from the Ozone Mapping Profiler Suite (OMPS; https://ozoneaq.gsfc.nasa.gov/, checked January 25, 2017), which has supplemented the earlier Total Ozone Mapping Spectrometer (TOMS) sensor, measures and reports the depletion of the stratospheric ozone layer over Earth's Poles, confirming previously hypothesized levels of ozone, and its declination (Molina and Rowland 1974; De Leeuw et al. 2010; Engel-Cox and Hoff 2005). At present, remote sensing technology is used to track global ozone layer holes on a daily basis by a partnership between the U.S. National Oceanic and Atmospheric Administration and the National Aeronautics and Space Administration (NASA 2011), and these data need to continue so we may monitor this important global phenomenon into the foreseeable future, given the critical importance of the issue of stratospheric ozone depletion.

The initiation of large-scale programs such as the Global Monitoring for Environment and Security (GMES; http://www.esa.int/About_Us/Ministerial _Council_2012/Global_Monitoring_for_Environment_and_Security_GMES, checked January 25, 2017) for the European Union (Backaus and Beule 2005) and the National Ecological Observatory Network (NEON; http://www .neonscience.org/, checked January 25, 2017) in the United States also highlight the growing importance of remote sensing, spatial databases, and global scale ecological data sets. NEON makes heavy use of remote sensing information, using an airborne observation platform (supporting LiDAR, digital photography, and imaging spectrometer sensors) to observe regional trends in land use and land cover change, invasive species, and ecosystem functioning (Kampe et al. 2010). This effort is all in tandem with ongoing and critically important extensive in situ monitoring and data collection, all to leverage geospatial modeling for a better understanding of processes.

Another important development over the past decade is the Global Earth Observation System of Systems (GEOSS), briefly referred to earlier in this chapter, which endeavors to combine both remote sensing data and ground-based data into a meta-database that could be used to identify global environmental risks and evaluate management and policy successes at the global scale (Herold et al. 2008; Stone 2010), long term. GEOSS provides information and tools for nine areas of focus, worldwide (Christian 2005; Lautenbacher 2006; Group on Earth Observations 2011):

1. Disasters
2. Health
3. Energy
4. Climate
5. Agriculture
6. Ecosystems
7. Biodiversity
8. Water
9. Weather

The development of GEOSS is driven by the Group on Earth Observations (GEO), an international consortium with members from 85 countries and numerous international organizations (Stone 2010). GEOSS is one example of the growing movement to release large data sets to the public, such as NASA's release of all historic Landsat imagery and the European Space Agency's intent to make all future satellite imagery free and open access (Stone 2010). The innovative goals of the GEOSS is to create a system of systems and supply data and tools to a larger audience of the public and decision makers, and this global effort is a tremendous advancement toward full access of remote sensing data and landscape-level products to all of those around the globe that could benefit from such information (http://www.earthobservations .org/geoss.php, checked January 25, 2017).

At this global scale, then, remote sensing data are beginning to form a more whole picture of the condition and status of ecosystems and, to a lesser degree, ecosystem functions and services. Such efforts will continue to lead us, as a global society, toward a much more detailed understanding of not just ecosystem conditions, such as vegetational community characteristics, hydrologic regimes, and soil conditions, but also provide us with a better integrated view of the globe, in terms of common trends and conditions amongst diverse and vast areas of land. The biophysical details of the work described in this book, and the projects in many other geographic locations around the world, are providing an excellent database for use in the effort to better understand our global condition and status, and to provide better early detection of potential degradation that may affect ecosystem functions at a broad scale, such as drought, fire, and sea level rise, and the associated global ecosystem services. This integration of local to regional to global analyses is powered by the technological capability of remote sensing and the tools of landscape ecology, allowing communities and decision makers of the future to take the most appropriate actions that will benefit all of society and the planet where we reside. To this end, remote sensing and landscape ecology will continue to contribute to humanity well into the future.

Glossary

ANOVA: Analysis of Variance test.

Anoxic: Condition which lacks oxygen, typical of wetland soils.

C-CAP: The U.S. National Atmospheric and Oceanographic Administration's Coastal Change and Analysis Program.

Decision Support: A set of software or database applications that are intended to allow users to search large amounts (e.g., in a clearinghouse) of information for specific reporting that can result in making (e.g., environmental) management decisions.

Ecological Driver: An ecological element that causes a change in an organism, community, ecosystem, or other ecological component of the landscape. An ecological driver may be biotic (e.g., an invasive plant species that causes a decrease in the biological diversity of a forest) or abiotic (e.g., a fire that causes a decrease in the biological diversity of a forest).

Ecological Goods and Services: See *Ecosystem Services*.

Ecological Indicator: A characteristic of the environment that is measured to provide evidence of the biological condition of a resource (Hunsaker and Carpenter 1990). Ecological indicators can be measured at different levels, including organism, population, community, or ecosystem. The indicators in this volume are measures of ecosystem level characteristics at a broad scale (Jones et al. 1997).

Ecological Processes: The flow of energy and nutrients (including water) through an ecosystem.

Ecological Receptor: An ecological element that is affected (either directly or indirectly) by an ecological driver. For example, the understory plant species that are extirpated as a result of a forest fire are ecological receptors.

Ecological Significance: Refers to a result or phenomenon that has relevance and applicability to ecological processes and elements (see *Statistical Significance*).

Ecology: The study of the interrelationships of the biotic and abiotic elements of the planet.

Ecosystem: An interacting system consisting of groups of organisms and their nonliving or physical environment, which are interrelated.

Ecosystem Approach: An approach to perceiving, managing, and otherwise living in an ecosystem that recognizes the need to preserve the ecosystem's biochemical pathways upon which life within the ecosystems depends (e.g., biological, social, economic, etc.).

Ecosystem Integrity: The inherent capability of an ecosystem to organize (e.g., its structures, processes, diversity) in the face of environmental change.

Ecosystem Services: A concept and state where the dynamic complex of plant, animal, and microorganism communities and the nonliving environment, interacting as a functional unit, include humans as an integral part of ecosystems (after MEA 2005). Synonymous with *Ecological Goods and Services.*

Endpoint: Describes a characteristic of an ecosystem of interest and should be an ecologically relevant measurement. An endpoint can be any parameter, from a biochemical state to an ecological community's functional condition.

Escherichia coli (E. coli): A fecal coliform type of bacteria that can be used to indicate water quality in streams and rivers.

Estuary: Wetland and other areas where a river meets the sea.

Eutrophic: Waters or soils that are rich in nutrients and have high primary productivity.

Extirpation: The elimination or disappearance of a species or subspecies from a particular area, but not from its entire range.

FGDC: The Federal Geographic Data Committee, which coordinates the development of the National Spatial Data Infrastructure.

Geospatial Statistical Techniques: A statistical analysis methodology that incorporates geographic parameters like shape, size, and configuration of landscape elements.

GEOSS: Global Earth Observation System of Systems.

GIS: Geographic Information System(s).

GLNPO: USEPA's Great Lakes National Program Office.

HGM: Hydrogeomorphic (methodology).

Hydraulic Model: Any model based upon the conveyance of liquids through pipes and channels, especially as a source of mechanical force or control.

Hydrologic Model: Any model based upon representations of a part of the earth's hydrologic cycle, which are typically used for the predictions.

Hyper-Eutrophication: The undesirable overgrowth of vegetation and algae as a result of high concentrations of nutrients in wetlands; eutrophication greater than the typically higher levels of nutrients found in wetland relative to lakes, streams, and rivers.

Hyperspectral Data: A remote sensing data type that contains a relatively large number of spectral bands (typically more than 20) and is acquired by a sensor that resides on either an Earth-orbiting or airborne platform (see *Multispectral Data*).

Indicator: In biology/ecology, any biological or ecological entity that characterizes the presence or absence of specific environmental conditions, as demonstrated by statistical correlations of ecologically meaningful relationships between the entity(ies) and the environmental condition(s).

Indirect Environmental Changes: Changes in environmental conditions or states that result from diffuse or other variable sources, which may include human activities, weather, and other ambient conditions.

Karst Topography: A type of topography that is formed over limestone, dolomite, or gypsum by dissolution and characterized by sinkholes, caves, and underground drainage.

Kilometer: 0.62 miles.

Land Cover: A biological or physical description of the Earth's surface. It is that which overlays or currently covers the ground. This description enables various biophysical categories to be distinguished, such as areas of vegetation (trees, bushes, fields, lawns), bare soil, hard surfaces (rocks, buildings), and wet areas and bodies of water (watercourses, wetlands).

Land Use: A social or economic description of land cover. For example, an "urban" land cover description can be described as a land use if particular information about the activities that occur in the urban area can be discerned, such as residential, industrial, or commercial uses. It may be possible to infer land use from land cover, and the converse, but situations are often complicated, and the links to land use are not always evident; unlike land cover, land use is difficult to infer from remote sensing imagery, or over vast areas of the landscape. For example, it is often difficult to decide if grasslands are used or not for agricultural purposes. Distinctions between land use and land cover and their definition have impacts on the development of classification systems, data collection, and geographic information systems in general.

Landsat: The satellite-based U.S. National Aeronautics and Space Administration project that, in the late 1960s and early 1970s, endeavored to observe land features from space. The program has evolved by the launching of a total of eight satellites to date. Landsat imagery is used for a variety of Earth observations.

Landscape: An area where the traits, patterns, and structure of a specific geographic area, including its biological composition, physical environment, and anthropogenic or social patterns exist and are examined or described.

Landscape Characterization: The process of documenting the traits and patterns of the essential elements of the landscape.

Landscape Ecology: The study of the distribution patterns of communities and ecosystems, the ecological processes that affect those patterns, and changes in pattern and process over time and space.

Landscape Indicator: A measurement of the landscape, calculated from mapped or remotely sensed data, used to describe some other spatial or temporal pattern(s) of land use or land cover across a geographic area.

Landscape Metric: A measurement of a component or components (e.g., patches of forest) within the landscape, which is used to characterize composition and spatial configuration of the component within the landscape (e.g., forest size, fragmentation, proximity to other land cover types).

Landscape Unit: A reference unit (usually of area) that is being measured, mapped, or described.

Liter: 1.057 quarts.

Marsh: A wetland with primarily herbaceous vegetation that is typically flooded or moist much of the year.

Meter: 3.28 feet.

Metric: Any measurement value.

Model: A representation of reality used to simulate a process, understand a situation, predict an outcome, or analyze a problem. A model is structured as a set of rules and procedures, including spatial modeling tools that relate to locations on the Earth's surface (Jones et al. 1997).

MODIS: The satellite-based Moderate Imaging Spectroradiometer. A project undertaken by the U.S. National Aeronautics and Space Administration that endeavored to improve our understanding of global dynamics and processes occurring on the land, in the oceans, and in the lower atmosphere.

Multispectral Data: A remote sensing data type that contains a relatively small number of spectral bands (typically less than ten) and is acquired by a sensor that resides on either an Earth-orbiting or airborne platform (see *Hyperspectral Data*).

Nonpoint-Source Pollutant: Any contaminant that is caused by rainfall or snowmelt moving over and through the ground, often in a diffuse manner.

Normalized Differential Vegetation Index (NDVI): A commonly used vegetation index that provides simple graphical indicators of vegetation condition or presence from remote sensing data; the NDVI is based upon visible and near-infrared spectra and is utilized to detect the proportion of live green vegetation in the landscape. (See https://phenology.cr.usgs.gov/ndvi_foundation.php, checked January 25, 2017.)

Oligotrophic: Waters or soils that are poor in nutrients and have low primary productivity.

Orthorectification: The process of removing the effects of sensor "tilt" and "terrain" distortion effects in remote sensing imagery for the purpose of creating a planimetrically correct image.

Patch: A discrete land cover unit, for example, a "patch of forest" is a specific 25-acre wooded area in Taney County, Missouri.

Perforated: The condition of a patch where gaps in the patch exist, such as a gap in a forest patch, which may contain shrub, grass, or other non-forestland cover.

Point-Source Pollutant: Any contaminant that is caused by an outflow from a measurable source, such as a contaminant that flows out of a pipe or channel into the environment, such as a waterbody.

Pour Point: A location at which water exits a watershed, which can be used to build a geospatial model of a watershed "contributing area" above; used for the determination of contributing land cover or other landscape conditions as it pertains to the water quality or water quantity at that "pour point."

Primary Productivity: The rate at which biomass is produced by organisms, which synthesize complex organic substances from simple inorganic substrates, such as in photosynthesis.

PRISM: Parameter-elevation Regressions on Independent Slopes Model.

Quantile Method: For thematic mapping, each class contains an approximately equal number (count) of features. A quantile classification is well-suited to linearly distributed data. Because features are grouped by the number within each class, the resulting map can be misleading, in that similar features can be separated into adjacent classes, or features with widely different values can be lumped into the same class. This distortion can be minimized by increasing the number of classes.

Reference Condition: Any standard against which the current condition of a system is compared, which is a method used in determining the relative condition of ecosystems and landscapes, using metrics and indicators.

Reporting Unit: Any defined area (e.g., an 8-digit USGS hydrologic unit code "HUC" or portion thereof) for which a landscape metric (e.g., percent urban) is calculated.

Riparian: An area of land that is on, or related to, the banks of a stream or river; often a wetland area or otherwise moist/wet area of land adjacent to a stream or river.

RUSLE: Revised Universal Soil Loss Equation.

Scale: The spatial or temporal dimension over which an object or process can be said to exist as in, for example, the scale of forest habitat. This is an important factor to consider during landscape ecology assessments because measured values often change with the scale of measurement. For example, coarse scale maps have less detailed information than fine scale maps and thus exclude some information, relative to fine scale maps.

Seiche: A temporary displacement of water in a large lake as a result of high winds or atmospheric pressure. The short-term water-level oscillations that result from a seiche are functionally analogous to ocean tides.

Spatial Database: A collection of information that contains data on the phenomenon of interest, such as forest condition or stream pollution, and the location of the phenomenon on the Earth's surface (Jones et al. 1997).

Spatial Filtering: A remote sensing data processing technique whereby each pixel value is changed by some predetermined function that relates to other pixels in the imagery, typically neighboring pixels.

Spatial Pattern: Generally, the way things are arranged on the Earth's surface, and thus on maps. For example, the pattern of forest patches can be described by their number, size, shape, or proximity to other entities. The spatial pattern exhibited by a map can be described in terms of its overall texture, complexity, or by other landscape metrics.

Spatial Signature: The concept and ability to characterize an area of the landscape by noting, measuring, and analyzing spatial patterns (Frohn 1997).

Standard Deviation Method: For thematic mapping, classes show the amount a feature's attribute value varies from the mean value of the distribution. Class breaks are generated by successively adding or subtracting the standard deviation from the mean. A two-color ramp is best used to emphasize values above or below the mean. It is particularly useful in viewing spatial variability of a parameter.

Statistical Significance: Refers to a result or phenomenon that is at a level of probability that it is determined to be relevant to the analyses undertaken. (See *Ecological Significance*.)

STATSGO: State Soil Geographic (database).

Sustainability: The concept and state of the ability to be sustained, supported, or maintained and not being (net) harmful to the environment or depleting of natural resources.

Swamp: A wetland with primarily forested or woody/shrubby vegetation, which is typically flooded or moist much of the year.

System: An assemblage of interrelated elements or components that comprise a unified whole. An ecological system (ecosystem) is one type.

Thematic Map: A map that shows the spatial distribution of one or more specific "data themes" (e.g., percentage of agriculture or human population).

Tidal Flat: Often referred to as mudflats, these coastal wetlands form near the mouths of rivers and streams where sediment settles out of the water column. Tidal flats are a typical feature of estuaries and other coastal inlets.

Transdisciplinary: A conceptual approach for collaborative work, which involves the fullest possible inclusion of expertise needed to address all of the elements of a work topic.

USEPA: United States Environmental Protection Agency.

Watershed: A region or area shown in a map as a bounded area that might be actually bounded (on the ground) by ridge lines or other physical divides, which drain ultimately to a particular watercourse or body of water (Jones et al. 1997).

References

Abood, S., A. Maclean, and L. Mason. 2012. Modeling riparian zones utilizing DEMS and flood height data. *Photogrammetric Engineering and Remote Sensing* 78:259–269.

Ackleson, S., and V. Klemas. 1987. Remote sensing of submerged aquatic vegetation in lower Chesapeake Bay: A comparison of Landsat MSS to TM imagery. *Remote Sensing of Environment* 22:235–248.

Acreman, M., and J. Holden. 2013. How wetlands affect floods. *Wetlands* 33:773–786.

Adam, E., O. Mutanga, and D. Rugege. 2010. Multispectral and hyperspectral remote sensing for identification and mapping of wetland vegetation: A review. *Wetlands Ecology and Management* 18:281–296.

Agnon, Y., and M. Stiassnie. 1991. Remote sensing of the roughness of a fractal sea surface. *Journal of Geophysical Research* 96:12773–12779.

Akins, E., Y. Wang, and Y. Zhou. 2010. EO-1 Advanced Land Imager data in submerged aquatic vegetation mapping, in *Remote Sensing of Coastal Environment*, ed. Y. Wang. Boca Raton, FL: CRC Press.

Allen, A. 1987. Habitat suitability index models: Mallard (winter habitat, Lower Mississippi Valley). *Biological Report* 82(10.132). Washington, DC: Department of Interior, U.S. Fish and Wildlife Service.

Alley, R. B., J. Marotzke, W. D. Nordhaus, J. T. Overpeck, D. M. Peteet, R. A. Pielke, R. T. Pierrehumbert, P. B. Rhines, T. F. Stocker, L. D. Talley, and J. M. Wallace. 2003. Abrupt climate change. *Science* 299:2005–2010.

Anderson, J., E. Hardy, J. Roach, and R. Witmer. 1976. A land use classification system for us with remote-sensor data. U.S. Geological Survey Professional Paper 964. Washington, DC: U.S. Department of Interior.

Antolovich, J. 2011. Disaster response aerial remote sensing following storms Irene and Lee. *Photogrammetric Engineering and Remote Sensing* 77:1185–1187.

Arkansas GAP Analysis Project. 2008. AR-GAP webpage. Available at http://web .cast.uark.edu/gap/data.html (checked January 25, 2017). Fayetteville, AR: University of Arkansas.

ASME. 2008. ASME water management technology vision and roadmap: Executive summary. Washington, DC: American Society of Mechanical Engineers.

Asner, G. P. 2011. Painting the world REDD: Addressing scientific barriers to monitoring emissions from tropical forests. *Environmental Research Letters* 6(2):021002.

Asner, G. P., D. E. Knappa, T. Kennedy-Bowdoina, M. O. Jones, R. E. Martina, J. Boardmanb, and R. F. Hughes. 2008. Invasive species detection in Hawaiian rainforests using airborne imaging spectroscopy and LiDAR. *Remote Sensing of Environment* 112:942–1955.

Asner, G. P., R. E. Martin, D. E. Knapp, R. Tupayachi, C. Anderson, L. Carranza, P. Martinez, M. Houcheime, F. Sinca, P. Weiss. 2011. Spectroscopy of canopy chemicals in humid tropical forests. *Remote Sensing of Environment* 115(12):3587–98.

Asner, G. P., P. G. Brodrick, C. B. Anderson, N. Vaughn, D. E. Knapp, and R. E. Martin. 2016. Progressive forest canopy water loss during the 2012–2015 California drought. *Proceedings of the National Academy of Sciences* 113:E249–E255.

Backaus, R., and B. Beule. 2005. Effective evaluation of satellite data products in environmental policy. *Space Policy* 21:173–183.

Bailey, R. G. 2009. *Ecosystem Geography: From Ecoregions to Sites*, 2nd ed. New York: Springer.

Baker, C., R. Lawrence, C. Montagne, and D. Patten. 2006. Mapping wetlands and riparian areas using Landsat ETM+ imagery and decision-tree–based models. *Wetlands* 26:465–474.

Baker, J. P., D. W. Hulse, S. V. Gregory, D. White, J. Van Sickle, P. A. Berger, D. Dole, and N. H. Schumaker. 2004. Alternative futures for the Willamette River basin, Oregon. *Ecological Applications* 14(2):313–324.

Baker, W. L., and Y. Cai. 1992. The r. le programs for multiscale analysis of landscape structure using the GRASS geographical information system. *Landscape Ecology* 7:291–302.

Ball, H., J. Jalava, T. King, L. Maynard, B. Potter, and T. Pulfer. 2003. *The Ontario Great Lakes Coastal Wetland Atlas*. Canada: Environment Canada and Ontario Ministry of Natural Resources.

Balzarolo, M., S. Vicca, A. L. Nguy-Robertson, D. Bonal, J. A. Elbers, Y. H. Fu, T. Grünwald, J. A. Horemans, D. Papale, J. Peñuelas, and A. Suyker. 2016. Matching the phenology of Net Ecosystem Exchange and vegetation indices estimated with MODIS and FLUXNET in-situ observations. *Remote Sensing of Environment* 174:290–300.

Barnard, P. L., and D. Hoover. 2010. A seamless, high-resolution, coastal digital elevation model (DEM) for southern California. Reston, VA: U.S. Geological Survey Data Series 487.

Bastin, L., and C. Thomas. 1999. The distribution of plant species in urban vegetation fragments. *Landscape Ecology* 14:493–507.

Benavides, J. A., B. Pietruszewski, B. Kirsch, and P. Bedient. 2001. Analyzing flood control alternatives for the Clear Creek watershed in a geographic information systems framework. In World Water and Environmental Resources Congress 2001. Orlando, FL: American Society of Civil Engineers.

Bengtson, M. L., and G. Padmanabhan. 1999. Hydrologic model for assessing the influence of wetlands on flood hydrographs in the Red River basin—Development and application. Fargo, ND: North Dakota Water Resources Research Institute, North Dakota State University.

Benniger, L. D. 1985 (April 30). Fallout PU and Natural U and PD in sediments of the north river marsh, North Carolina. EOS Transactions, American Geophysical Union, 66, 18.

Betbedera, J., L. Hubert-Moya, F. Burelb, S. Corgnea, and J. Baudryc. 2015. Assessing ecological habitat structure from local to landscape scales using synthetic aperture radar. *Ecological Indicators* 52:545–557.

Blom, C., G. Bogemann, P. Lann, A. Van der Sman, H. Van der Steeg, and L. Voesenek. 1990. Adaptations to flooding in plants from river areas. *Aquatic Botany* 38:29–47.

Bolstad, P., and T. Lillesand. 1992. Rule-based classification models: Flexible integration of satellite imagery and thematic spatial data. *Photogrammetric Engineering and Remote Sensing* 58:965–971.

Bormann, F. H., G. E. Likens, D. W. Fisher, and R. S. Pierce. 1968. Nutrient loss accelerated by clear-cutting of a forest ecosystem. *Science* 159:882–884.

Bourgeau-Chavez, L., S. Endres, M. Battaglia, M. E. Miller, E. Banda, Z. Laubach, P. Higman, P. Chow-Fraser, and J. Marcaccio. 2015. Development of a bi-national Great Lakes coastal wetland and land use map using three-season PALSAR and Landsat imagery. *Remote Sensing* 7:8655–8682.

Bromberg, S. 1990. Identifying ecological indicators: An environmental monitoring and assessment program. *Journal of the Air Pollution Control Association* 40:976–978.

Brown, M., and J. Dinsmore. 1986. Implications of marsh size and isolation for marsh bird management. *Journal of Wildlife Management* 50:392–397.

Brown, D., E. Addink, J. Duh, and M. Bowersox. 2004. Assessing uncertainty in spatial landscape metrics derived from remote sensing data, in *Remote Sensing and GIS Accuracy Assessment*, eds. R. S. Lunetta and J. G. Lyon. Boca Raton, FL: CRC Press.

Brown, D. E., C. H. Lowe, and C. P. Pase. 1979. A digitized classification system for the biotic communities of North America, with community (series) and association examples for the Southwest. *Journal of the Arizona-Nevada Academy of Science* 14:1–16.

Bryan, B. A., S. Hajkowicz, S. Marvanek, and M. D. Young. 2009. Mapping economic returns to agricultural for informing environmental policy in the Murray–Darling Basin, Australia. *Environmental Modeling and Assessment* 14:375–390.

Burnicki, A. 2011. Modeling the probability of misclassification in a map of land cover change. *Photogrammetric Engineering and Remote Sensing* 77:39–49.

Butera, K. 1983. Remote sensing of wetlands. *IEEE Transactions on Geoscience and Remote Sensing* GE-21:383–392.

Buzzelli, C. P. 2009. Measurement of water colour using AVIRIS imagery to assess the potential for an operational monitoring capability in the Pamlico Sound estuary, U.S.A. *International Journal of Remote Sensing* 30:3291–3314.

Cahoon, D., J. Day, and D. Reed. 1999. The influence of surface and shallow subsurface soil processes on wetland elevation: A synthesis. *Current Topics in Wetland Biogeochemistry* 3:72–88.

Callan, O., and A. Mark. 2008. Using MODIS data to characterize seasonal inundation patterns in the Florida Everglades. *Remote Sensing of Environment* 112:4107–4119.

Carter, V. 1990. Importance of hydrologic data for interpreting wetland maps and assessing wetland loss and mitigation. Biology Report 90(18). Washington, DC: Department of Interior, U.S. Fish and Wildlife Service.

Cayan, D., P. Bromirski, K. Hayhoe, M. Tyearee, M. Dettinger, and R. Flick. 2006 (March). Projecting Future Sea Level White Paper. Sacramento, CA: California Climate Change Center.

Chander, G., B. Markham, and D. Helder. 2009. Summary of current radiometric calibration coefficients for Landsat MSS, TM, ETM+, and EO-1 ALI sensors. *Remote Sensing of Environment* 113:893–903.

Changnon, S. 2008. Assessment of flood losses in the United States. *Journal of Contemporary Water Research and Education* 138:38–44.

Chirico, J., and G. S. Farley (eds.). 2015. *Thinking Like and Island: Navigating a Sustainable Future in Hawai'i*. Honolulu, HI: University of Hawai'i Press.

Cho, M. A., R. Mathieua, G. P. Asner, L. Naidooa, J. van Aardt, A. Ramoeloa, D. Pravesh, K. Wesselse, M. Russell, I. P. J. Smitf, and B. Erasmusg. 2012. Mapping tree species composition in South African savannas using an integrated airborne spectral and LiDAR system. *Remote Sensing of Environment* 125:214–226.

Christian, E. 2005. Planning for the Global Earth Observation System of Systems (GEOSS). *Space Policy* 21:105–109.

Clough, J., and C. Larson. 2010. SLAMM 6.0.1, User's Manual. Warren Pinnacle Consulting, Inc.

Clough, J., R. Park, and R. Fuller. 2010. SLAMM 6 Technical Documentation, Release 6.0.1. Warren Pinnacle Consulting, Inc.

Combs, J. L., and A. C. Perry. 2003. The 1903 and 1993 Floods in Kansas—The Effects of Changing Times and Technology. USGS Fact Sheet 019-03.

Conagua. 2015. *Atlas del Agua en México*. Semarnat-Gobierno Federal, México.

Congalton, R., and K. Green. 2009. *Accuracy Assessment of Remotely Sensed Data: Principles and Practices*, 2nd ed. Boca Raton, FL: CRC/Lewis Publishers.

Connell, J., and R. Slatyer. 1977. Mechanisms of succession in natural communities and their role in community stability and organization. *American Naturalist* 111:1119–1144.

Cooper, M. J. P., M. D. Beevers, and M. Oppenheimer. 2008. The potential impacts of sea level rise on the coastal region of New Jersey, USA. *Climate Change* 90:475–492.

Costanza, R. 1980. Embodied energy and economic evaluation. *Science* 210:1219–1224.

Costlow, J., C. Boakout, and R. Monroe. 1960. The effect of salinity and temperature on larval development of *Sesarma cincereum* (Bosc.) reared in the laboratory. *Biological Bulletin* 118:183–202.

Cowardin, L., V. Carter, F. Gollet, and E. LaRoe. 1979. Classification of Wetlands and Deepwater Habitats of the United States. FWS/OBS 79/31. Washington, DC: Department of Interior, U.S. Fish and Wildlife Service.

Cox, K. W., and G. Cintrón. 1997. The North American Region in Wetlands, Biodiversity and the Ramsar Convention, in Wetlands, Biodiversity and the Ramsar Convention: The Role of the Convention on Wetlands in the Conservation and Wise Use of Biodiversity, ed. A.J. Hails. Ramsar Convention Bureau, Gland, Switzerland. Accessed from http://www.ramsar.org/sites/default/files/documents/library/wetlands_biodiversity_and_the_ramsar_convention.pdf (checked January 25, 2017).

Crumpton, W. G., G. A. Stenback, B. A. Miller, and M. J. Helmers. 2006. Potential Benefits of Wetland Filters for Tile Drainage Systems: Impact on Nitrate Loads to Mississippi River Subbasins. Final Project Report, project No. IOW06682. Washington, DC: U.S. Department of Agriculture.

Cushman, S., and K. McGarigal. 2004. Patterns in the species-environmental relationship depend on both scale and choice of response variables. *Oikos* 105:117–124.

Dahl, T. 1990. Wetlands Losses in the United States, 1780s to 1980s. Washington, DC: Department of Interior, U.S. Fish and Wildlife Service.

Dahl, T. 2006. Status and Trends of Wetlands in the Conterminous United States 1998 to 2004. Washington, DC: Department of Interior, U.S. Fish and Wildlife Service.

Dahl, T., and C. Johnson. 1991. Status and Trends of Wetlands in the Conterminous United States, Mid-1970s to Mid-1980s. Washington, DC: Department of Interior, U.S. Fish and Wildlife Service.

Dahl, T., and M. Watmough. 2007. Current approaches to wetland status and trends monitoring in prairie Canada and the continental United States of America. *Canadian Journal of Remote Sensing* 33:S17–S27.

Dale, V. H., R. V. O'Neill, M. Pedlowski, and F. Southworth. 1993. Causes and effects of land use change in central Rondônia, Brazil. *Photogrammetric Engineering and Remote Sensing* 59:997–1005.

Dale, V. H., R. V. O'Neill, F. Southworth, and M. Pedlowski. 1994. Modeling in the Brazilian Amazonian settlement of Rondônia. *Conservation Biology* 8:196–206.

De Cola, L. 1989. Fractal analysis of a classified Landsat scene. *Photogrammetric Engineering and Remote Sensing* 55:601–610.

De Leeuw, J., Y. Georgiadou, N. Kerle, A. de Gier, Y. Inoue, J. Ferwerda, M. Smies, and D. Narantuya. 2010. The function of remote sensing in support of environmental policy. *Remote Sensing* 2:1731–1750.

Diamond, J. 1974. Colonization of exploded volcanic islands by birds: The super tramp strategy. *Science* 184:803–806.

Doyle, M., and C. A. Drew. 2008. *Large-Scale Ecosystem Restoration: Five Case Studies from the United States.* Washington, DC: Island Press.

Dzwonko, Z., and S. Loster. 1988. Species richness of small woodlands on the western Carpathian foothills. *Vegetatio* 76:15–27.

Ehman, J. 2008. Data Report on SLAMM Model Results for Ten National Wildlife Refuges in South Carolina and Georgia: Wolf Island NWR, Georgia. Image Matters LLC.

Ehrenfeld, J. 1983. The effects of changes in land-use on swamps of the New Jersey Pine Barrens. *Biological Conservation* 25:253–275.

Ehrenfeld, J., and J. Schneider. 1991. *Chamaecyparis thyoides* wetlands and suburbanization: Effects on hydrology, water quality and plant community composition. *Journal of Applied Ecology* 28:467–490.

Engel-Cox, J. A., and R. M. Hoff. 2005. Science-policy data compact: Use of environmental monitoring data for air quality policy. *Environmental Science Policy* 8:115–131.

Environment Canada. 1995. Great Lakes Fact Sheet: Amphibians and Reptiles in Great Lakes Wetlands: Threats and Conservation. Canadian Wildlife Service. Accessed from https://brocku.ca/massasauga/bibliography/Shirose%20et%20al%201995.pdf (checked January 25, 2017).

Environment Canada. 1998. Great Lakes Fact Sheet: How Much Habitat Is Enough? Minister of Public Works and Government Services, Canada.

Environment Canada. 2002. Great Lakes Fact Sheet: Great Lakes Coastal Wetlands— Science and Conservation. Canadian Wildlife Service (Ontario Region).

Falkner, E., and D. D. Morgan. 2001. *Aerial Mapping: Methods and Applications*, 2nd ed. Boca Raton, FL: Lewis Publishers/CRC Press.

Federal Geographic Data Committee. 1992. Application of Satellite Data for Mapping and Monitoring Wetlands. Washington, DC: Wetlands Subcommittee.

Federal Geographic Data Committee. 2008. Federal Geographic Data Committee National Vegetation Classification. FGDC-STD-005-2008 (Version 2). Federal Geographic Data Committee, US Geological Survey, Reston, Virginia, USA. http://www. fgdc. gov/standards/projects/FGDC-standardsprojects/vegetation/standards/projects/vegetation.

Federal Geographic Data Committee. 2010. Geospatial Metadata: Federal Geographic Data Committee Website. Accessed from http://www.fgdc.gov/metadata/geospatial-metadata-standards (checked January 25, 2017).

FEMA and the State of North Carolina. 2003. A Report of Flood Hazards in Carteret County, North Carolina, and Incorporated Areas. Flood Insurance Study Number 37031CV000A.

Fenga, Y., and Y. Liu. 2015. Fractal dimension as an indicator for quantifying the effects of changing spatial scales on landscape metrics. *Ecological Indicators* 53:18–27.

Foody, G. 2002. Status of landcover classification accuracy assessment. *Remote Sensing of the Environment* 80:185–201.

Forman, R. 1995. *Land Mosaics: The Ecology of Landscapes and Regions.* Cambridge, UK: Cambridge University Press.

Forman, R., and M. Godron. 1986. *Landscape Ecology.* New York: John Wiley & Sons.

Forman, R., A. Galli, and C. Leck. 1976. Forest size and avian diversity in New Jersey woodlots with some land use implication. *Oecologia* 26:1–8.

Frohn, R. 1997. *Remote Sensing for Landscape Ecology: New Metric Indicators for Monitoring, Modeling, and Assessment of Ecosystems.* Boca Raton, Florida: CRC Press.

Frohn, R., R. Molly, C. Lane, and B. Autrey. 2009. Satellite remote sensing of isolated wetlands using object-oriented classification of Landsat-7 data. *Wetlands* 29:931–941.

Gardner, R. H., B. T. Milne, M. Turner, and R. V. O'Neill. 1987. Neutral models for the analysis of broad-scale landscape pattern. *Landscape Ecology* 1:19–28.

Garofalo, D. 2003. Aerial photointerpretation of hazardous waste sites: An overview, in *Geographic Information System Applications for Watershed and Water Resources Management*, ed. J. G. Lyon. London: Taylor & Francis.

Gesch, D. B. 2009. Analysis of LiDAR elevation data for improved identification and delineation of lands vulnerable to sea-level rise. *Journal of Coastal Research* (Special Issue 53):49–58.

Gilmore, M., D. Civco, E. Wilson, N. Barrett, S. Prisloe, J. Hurd, and C. Chadwick. 2009. Remote sensing and in situ measurements for delineation and assessment of coastal marshes and their constituent species, in *Remote Sensing of Coastal Environments*, ed. J. Wang. Boca Raton, FL: CRC Press.

GLCWC (Great Lakes Coastal Wetlands Consortium). 2004a. Great Lakes Coastal Wetlands: Consortium Fact Sheet. Accessed from http://glc.org/projects/habitat/coastal-wetlands/ (checked January 25, 2017).

GLCWC (Great Lakes Coastal Wetlands Consortium). 2004b. Study Indicators and Metrics. Accessed from http://glc.org/files/projects/cwc/CWC-StudyIndicators-2001.pdf (checked January 25, 2017).

Glick, P. 2008. Sea-Level Rise and Coastal Habitats in the Chesapeake Bay Region Technical report. National Wildlife Federation.

GLIN (Great Lakes Information Network). 2004. The Great Lakes. Accessed from http://www.great-lakes.net/lakes/ (checked January 25, 2017).

GLNPO (Great Lakes National Program Office). 1999. Selection of indicators for Great Lakes basin ecosystem health, in ed. P. Bertram, version 3. State of the Lakes Ecosystem Conference.

Gonzaleza, P., G. P. Asner, J. J. Battlesa, M. A. Lefskyd, K. M. Waringa, and M. Palacee. 2010. Forest carbon densities and uncertainties from Lidar, QuickBird, and field measurements in California. *Remote Sensing of Environment* 114:1561–1575.

Goodchild, M. F., and D. M. Mark. 1987. The fractal nature of geographic phenomena. *Annals of the Association of American Geographers* 77:265–278.

Gorham, E. 1987. The natural and anthropogenic acidification of peatlands, in *Effects of Atmospheric Pollutants on Forests, Wetlands, and Agricultural Ecosystems*, eds. T. C. Hutchinson and K. M. Meema. Berlin, Germany: Springer-Verlag, 493–512.

Government of Canada and GLNPO (Great Lakes National Program Office). 1995. *The Great Lakes: An Environmental and Resource Book*, 3rd edition. Government of Canada, Toronto, Ontario, and U.S. Environmental Protection Agency, Great Lakes National Program Office, Chicago, IL.

Graham, R. L., C. T. Hunsaker, R. V. O'Neill, and B. Jackson. 1991. Ecological risk assessment at the regional scale. *Ecological Applications* 1:196–206.

Green, R. 1979. *Sampling Design and Statistical Methods for Environmental Biologists.* New York: John Wiley & Sons.

Group on Earth Observations (GEO). 2011. Geo Portal: GEOSS online. Accessed from http://www.earthobservations.org/index.html (checked January 25, 2017).

Gustafson, E. 1998. Quantifying landscape spatial pattern: What is the state of the art? *Ecosystems* 1:143–156.

Gustafson, E. J., and G. R. Parker. 1992. Relationships between landcover proportion and indices of landscape spatial pattern. *Landscape Ecology* 7:101–110.

Gutzwiller, K., and S. Anderson. 1992. Interception of moving organisms; influences of path shape, size, and orientation on community structure. *Landscape Ecology* 6:293–303.

Haggard, B. E., P. A. Moore, Jr., I. Chaubey, and E. H. Stanley. 2003. Nitrogen and phosphorus concentrations and export from an Ozark Plateau catchment in the United States. *Biosystems Engineering* 86:75–85.

Hamazaki, T. 1996. Effects of patch shape on the number of organisms. *Landscape Ecology* 11:299–306.

Harris, L. 1984. *The Fragmented Forest: Island Biogeography Theory and the Preservation of Biotic Diversity.* Chicago: University of Chicago Press.

Heberger, M., H. Cooley, P. Herrera, P. Gleick, and E. Moore. 2009. The Impacts of Sea-Level Rise on the California Coast. The California Climate Change Center, Pacific Institute. Accessed from http://pacinst.org/app/uploads/2014/04/sea-level-rise.pdf (checked January 25, 2017).

HEC. 2010a. HEC–HMS. Hydrologic Engineering Center: U.S. Army Corps of Engineers. Accessed from http://www.hec.usace.army.mil/software/hec-hms/ (checked January 25, 2017).

HEC. 2010b. HEC–GeoHMS. Hydrologic Engineering Center: U.S. Army Corps of Engineers. Accessed from http://www.hec.usace.army.mil/software/hec-geohms/index.html (checked January 25, 2017).

HEC. 2010c. HEC–HMS User's Manual Version 3.5. Hydrologic Engineering Center: U.S. Army Corps of Engineers.

HEC. 2010d. HEC–RAS User's Manual Version 4.1. Hydrologic Engineering Center: U.S. Army Corps of Engineers.

HEC. 2010e. HEC–GeoRAS. Hydrologic Engineering Center: U.S. Army Corps of Engineers. Accessed from http://www.hec.usace.army.mil/software/hec-georas/ (checked January 25, 2017).

Henderson, F., and A. Lewis. 2008. Radar detection of wetland ecosystems: A review. *International Journal of Remote Sensing* 29:5809–5835.

Hermy, M., and H. Stieperaere. 1981. An indirect gradient analysis of the ecological relationships between ancient and recent riverine woodlands to the south of Bruges (Flanders, Belgium). *Vegetatio* 44:43–49.

Herold, M., C. E. Woodcock, T. R. Loveland, J. Townshend, M. Brady, C. Steenmans, and C. C. Schmullius. 2008. Land-cover observations as part of a Global Earth Observation System of Systems (GEOSS): Progress, activities, and prospects. *IEEE Systems Journal* 2:414–423.

Hershfield, D. M. 1961. Rainfall Frequency Atlas of the United States for Durations from 30 Minutes to 24 Hours and Return Periods from 1 to 100 Years. Technical Paper No. 40. Washington, DC: U.S. Department of Commerce, Weather Bureau.

Hey, D., and N. Phillippi, 1995. Flood reduction through wetland restoration: The upper Mississippi as a case study. *Restoration Ecology* 3: 4–17.

Hinckley, E.-L. S., G. B. Bonan, G. J. Bowen, B. P. Colman, P. A. Duffy, C. L. Goodale, B. Z. Houlton, E. Marín-Spiotta, K. Ogle, S. V. Ollinger, E. A. Paul, P. M. Vitousek, K. C. Weathers, and D. G. Williams. 2016. The soil and plant biogeochemistry sampling design for the National Ecological Observatory Network. *Ecosphere* 7:e01234.10.1002/ecs2.1234.

Hirsch Hadorn, G., D. Bradley, C. Pohl, S. Rist, and U. Wiesmann. 2006. Implications of transdisciplinarity for sustainability research. *Ecological Economics* 60:119–128.

Homer, C., J. Dewitz, L. Yang, S. Jin, P. Danielson, G. Xian, J. Coulston, N. Herold, J. Wickham, and K. Megown. 2015. Completion of the 2011 National Land Cover Database for the Conterminous United States—Representing a decade of land cover change information. *Photogrammetric Engineering and Remote Sensing* 81:345–354.

Honkavaara, E., P. Litkey, and K. Nurminen. 2013. Automatic storm damage detection in forests using high-altitude photogrammetric imagery. *Remote Sensing* 5:1405–1424.

Houghton, J. T., Y. Ding, D. J. Griggs, M. Noguer, P. J. van der Linden, X. Dai, K. Maskell, and C. A. Johnson (eds). 2001. *Climate Change 2001: The Scientific Basis*. Contribution of Working Group I to the Third Assessment Report of the Intergovernmental Panel on Climate Change. Cambridge, UK: Cambridge University Press.

Howard, J. 1970. *Aerial Photo-Ecology*. New York: American Elsevier.

Huang, P., Y. Li, and M. Sumner. 2011. *Handbook of Soil Sciences*, 2nd ed. Boca Raton, FL: CRC Press.

Hunsaker, C., and D. Carpenter. 1990. *Environmental Monitoring and Assessment Program—Ecological Indicators*. EPA 600/3-90/060. Research Triangle Park, North Carolina: U. S. Environmental Protection Agency.

ICLUS. 2010. Integrated Climate and Land Use Scenarios (ICLUS) V1.3 User's Manual: ArcGIS Tools and Datasets for Modeling US Housing Density Growth. Global Change Research Program: U. S. Environmental Protection Agency.

IPCC. 2001. *Climate Change 2001: The Scientific Basis*. Contribution of Working Group I to the Third Assessment Report of the Intergovernmental Panel on Climate Change [J. T. Houghton, Y. Ding, D. J. Griggs, M. Noguer, P. J. van der Linden, X. Dai, K. Maskell, and C. A. Johnson (eds.)]. Cambridge University Press, Cambridge, United Kingdom and New York, NY, 881 pp.

Islam, M., P. Thenkabail, R. Kulawardana, R. Alankara, S. Gunasinghe, C. Edussriya, and A. Gunawardana. 2008. Semi-automated methods for mapping wetlands using Landsat ETM+ and SRTM data. *International Journal of Remote Sensing* 29:7077–7106.

Iverson, L. R. 1988. Land use changes in Illinois, USA: The influence of landscape attributes on current and historic land use. *Landscape Ecology* 2:45–61.

Jenning, M. 1995. Gap analysis today: A confluence of biology, ecology, and geography for management of biological resources. *Wildlife Society Bulletin* 23:658–662.

Jensen, J. 2004. *Introductory Digital Image Processing: A Remote Sensing Perspective.* Englewood Cliffs, NJ: Prentice Hall.

Johansen, K., L. Arroyo, S. Phinn, and C. Witte. 2010. Comparison of geo-object base and pixel-based change detection of riparian environments using high spatial resolution multi-spectral imagery. *Photogrammetric Engineering and Remote Sensing* 76:123–136.

Johnston, C. 1989. Human impacts to Minnesota wetlands. *Journal of the Minnesota Academy of Science* 55:120–124.

Johnston, C. 1994. Cumulative impacts to wetlands. *Wetlands* 14:49–55.

Jones, J. R., M. F. Knowlton, D. V. Obrecht, and E. A. Cook. 2004. Importance of landscape variables and morphology on nutrients in Missouri reservoirs. *Canadian Journal of Fisheries and Aquatic Sciences* 61:1503–1512.

Jones, K., A. Neale, M. Nash, R. Van Remortel, J. Wickham, K. Riitters, and R. O'Neill. 2000. Landscape correlates of breeding bird richness across the United States Mid-Atlantic Region. *Environmental Monitoring and Assessment* 63:159–174.

Jones, K., A. Neale, M. Nash, R. Van Remortel, J. Wickham, K. Riitters, and R. O'Neill. 2001. Predicting nutrient and sediment loadings to streams from landscape metrics: A multiple watershed study from the United State Mid-Atlantic Region. *Landscape Ecology* 16:301–312.

Jones, K., K. Riiters, J. Wickham, R. Tankersley, R. O'Neill, D. Chaloud, E. Smith, and A. Neale. 1997. An ecological assessment of the United States Mid-Atlantic Region: A landscape atlas. EPA/600/R-97/130. U.S. Environmental Protection Agency, Office of Research and Development.

Juliano, K., and S. P. Simonovic. 1999. *The Impact of Wetlands on Flood Control in the Red River Valley of Manitoba.* Winnipeg, Manitoba: University of Manitoba.

Juracek, F., A. Perry, and E. Putnam. 2001. The 1951 Floods in Kansas Revisited. USGS Fact Sheet 041-01. U.S. Geological Survey.

Kalkhan, M. A., E. J. Stafford, and T. J. Stohlgren. 2007. Rapid plant diversity assessment using a pixel nested plot design: A case study in Beaver Meadows, Rocky Mountain National Park, Colorado, USA. *Diversity and Distributions* 13:379–388.

Kampe, T. U., B. R. Johnson, M. Kuester, and M. Keller. 2010. NEON: The first continental-scale ecological observatory with airborne remote sensing of vegetation canopy biochemistry and structure. *Journal of Applied Remote Sensing* 4:043510.

Karr, J., and E. Chu. 1997. Biological Monitoring and Assessment: Using Multimetric Indexed Effectively. EPA/235/R97/001. Seattle, WA: University of Washington.

Keddy, P. A., H. T. Lee, and I. C. Wisheu. 1993. Choosing indicators of ecosystem integrity: Wetlands as a model system, in *Ecological Integrity and the Management of Ecosystems,* eds. S. Woodley et al. Delray Beach, FL: St. Lucie Press, 61–82.

Kellman, M. 1996. Redefining roles: Plant community reorganization and species preservation in fragmented systems. *Global Ecology and Biogeography Letters* 5:111–116.

Keough, J., T. Thompson, G. Guntenspergen, and D. Wilcox. 1999. Hydrogeomorphic factors and ecosystem responses in coastal wetlands of the Great Lakes. *Wetlands* 19:821–834.

KGS. 1998. The Kansas River Corridor—Its Geologic Setting, Land Use, Economic Geology, and Hydrology. Kansas Geological Survey. Accessed from http://www.kgs.ku.edu/Publications/KR/index.html (checked January 25, 2017).

Klemas, V. 2009. Sensors and techniques for observing coastal ecosystems, in *Remote Sensing and Geospatial Technologies for Coastal Ecosystem Assessment and Management*, ed. X. Yang. Berlin, Germany: Springer-Verlag, 17–44.

Klemas, V. 2011. Remote sensing of wetlands: Case studies comparing practical techniques. *Journal of Coastal Research* 27:418–427.

Knebl, M. R., Z.-L. Yang, K. Hutchison, and D. R. Maidment. 2005. Regional scale flood modeling using NEXRAD rainfall, GIS, and HEC–HMS/RAS: A case study for the San Antonio River Basin Summer 2002 storm event. *Journal of Environmental Management* 75:325–336.

Koren, I., L. A. Remer, and K. Longo. 2007. Reversal of trend of biomass burning in the Amazon. *Geophysical Research Letters* 34:L20404.

Krummel, J. R., R. H. Gardner, G. Sugihara, R. V. O'Neill, and P. R. Coleman. 1987. Landscape patterns in a disturbed environment. *Oikos* 48:321–324.

Lake Huron Centre. 2000. *Critical Ecosystems along Lake Huron: Coastal Wetlands*. Blyth, Ontario: Lake Huron Centre for Coastal Conservation. Accessed from http://lakehuron.ca/index.php?page=critical-ecosystems-along-lake-huron (checked January 25, 2017).

Lal, R., and B. Stewart. 2012. *World Soil Resources and Food Security*. Boca Raton, FL: CRC Press.

Lam, N. S. N. 1990. Description and measurement of Landsat TM images using fractals. *Photogrammetric Engineering and Remote Sensing* 56:187–195.

Lam, N. S. N., and D. Quattrochi. 1992. On the issues of scale, resolution, and fractal analysis in the mapping sciences. *Professional Geographer* 44:88–98.

Lang, M., E. Kasischke, S. Prince, and K. Pittman. 2008. Assessment of C-band synthetic aperture radar data for mapping and monitoring coastal plain forested wetlands in the Mid-Atlantic region, USA. *Remote Sensing of Environment* 112:4120–4130.

Lathrop, R. G., and D. L. Peterson. 1992. Identifying structural self-similarity in mountainous landscapes. *Landscape Ecology* 6:233–238.

Lautenbacher, C. C. 2006. The Global Earth Observation System of Systems: Science serving society. *Space Policy* 22:8–11.

Lee, C., and S. Marsh. 1995. The use of archival Landsat MSS and ancillary data in a GIS environment to map historical change in an urban riparian habitat. *Photogrammetric Engineering and Remote Sensing* 61:999–1008.

Lewis, W. 1995. *Wetlands Characteristics and Boundaries*. Washington, DC: National Academy Press.

Li, H., and J. F. Reynolds. 1993. A new contagion index to quantify spatial patterns of landscapes. *Landscape Ecology* 8:155–162.

Lillesand, T., R. Kiefer, and J. Chipman. 2014. *Remote Sensing and Image Interpretation*. New York: John Wiley & Sons.

Linderman, M., Y. Zeng, and P. Rowhani. 2010. Climate and land-use effects on interannual fAPAR variability from MODIS 250 m data. *Photogrammetric Engineering and Remote Sensing* 76:807–816.

Linnet, L. M., S. J. Clarke, C. Graham, and D. N. Langhorne. 1991. Remote sensing of the sea-bed using fractal techniques. *Electronics & Communication Engineering Journal* October:195–203.

Linsley, R., and J. Franzini. 1979. *Water Resources Engineering.* New York: McGraw-Hill.

Lopez, R. D. 1997. Seasonal Allocation of Zinc, Copper, Nickel, Lead, and Cadmium in Cattails (*Typha angustifolia*) in a Restored Lacustrine Coastal Wetland. Master's thesis. Columbus, OH: The Ohio State University.

Lopez, R. D. (ed.). In preparation. *Societal Dimensions of Environmental Sciences: Case Studies of Collaboration and Transformation.* Boca Raton, FL: CRC Press.

Lopez, R., C. Davis, and M. Fennessy. 2002. Ecological relationships between landscape change and plant guilds in depressional wetlands. *Landscape Ecology* 17:43–56.

Lopez, R., J. Lyon, L. Lyon, and D. Lopez. 2013. *Practical Tools, Methods, and Approaches for Landscape Ecology,* 2nd ed. Boca Raton, FL: CRC Press.

Lopez, R., D. Heggem, C. Edmonds, K. Jones, L. Bice, M. Hamilton, E. Evanson, C. Cross, and D. Ebert. 2003. A Landscape Case Study of Ecological Vulnerability: Arkansas' White River watershed and the Mississippi alluvial valley ecoregion. EPA/600/R-03/057. Washington, DC: U.S. Environmental Protection Agency.

Lopez, R., D. Heggem, D. Sutton, T. Ehil, R. Van Remortel, E. Evanson, and L. Bice. 2006a. Using Landscape Metrics to Develop Indicators of Great Lakes Coastal Wetland Condition. EPA/X-06/002. Las Vegas, NV: U.S. Environmental Protection Agency, Environmental Sciences Division.

Lopez, R., M. Nash, D. Heggem, L. Bice, E. Evanson, L. Woods, R. Van Remortel, M. Jackson, D. Ebert, and T. Harris. 2006b. Water Quality Vulnerability in the Ozarks Using Landscape Ecology Metrics: EPA/600/C-06/017. Washington, DC: U.S. Environmental Protection Agency.

Lory, J. 1999. Agricultural Phosphorus and Water Quality, Publication G9181. Columbia, MO: University of Missouri.

Loveland, T., and D. Ohlen, 1993. Experimental AVHRR land data sets for environmental monitoring and modeling, in *Environmental Modeling with GIS,* eds. M. Goodchild, B. Parks, and L. Steyaert. New York: Oxford University Press, 379–385.

Ludwig, A., W. C. Hession, D. Scott, and D. Gallagher. 2016. Simulated flood of a small constructed floodplain wetland in Virginia: Event-scale pollutant attenuation. *Transactions of the American Society of Agricultural and Biological Engineers* 59:1321–1331.

Lunetta, R., and C. Elvidge. 1998. *Remote Sensing Change Detection.* Chelsea, MI, and Boca Raton, FL: Ann Arbor Press/CRC Lewis Publishers.

Lunetta, R., J. Lyon, C. Elvidge, and B. Guindon. 1998. North American landscape characterization: Dataset development and data fusion issues. *Photogrammetric Engineering and Remote Sensing* 64:821–829.

Lunetta, R., Y. Shao, J. Ediriwickrema, and J. Lyon. 2010. Monitoring agricultural cropping patterns across the Laurentian Great Lakes Basin using MODIS-NDVI data. *International Journal of Applied Earth Observations and Geoinformation* 12:81–88.

Lunetta, R., J. Lyon, D. Worthy, J. Sturdevant, J. Dwyer, D. Yuan, C. Elvidge, and L. Fenstermaker. 1993. North American Landscape Characterization (NALC), Landsat Pathfinder technical plan. EPA 600/X-93/009. Las Vegas, NV: U.S. Environmental Protection Agency.

Lunetta, R. S., J. F. Knight, H. W. Paerl, J. J. Streicher, B. L. Peierls, T. Gallo, J. G. Lyon, T. H. Mace, and C. P. Buzzelli. 2009. Measurement of water colour using AVIRIS imagery to assess the potential for an operational monitoring capability in the Pamlico Sound Estuary, USA. *International Journal of Remote Sensing* 30(13):3291–3314.

Luo, X., X. Chen, L. Xu, R. Myneni, and Z. Zhu. 2013. Assessing performance of NDVI and NDVI3g in monitoring leaf unfolding dates of the deciduous broadleaf forest in Northern China. *Remote Sensing* 5:845–861.

Luoto, M. 2000. Modeling of rare plant species richness by landscape variable in an agricultural area in Finland. *Plant Ecology* 149:157–168.

Lyon, J. 1981. The Influence of Lake Michigan Water Levels on Wetland Soils and Distribution of Plants in the Straits of Mackinac, Michigan. Doctoral dissertation. Ann Arbor, MI: University of Michigan.

Lyon, J. 1987. Maps, aerial photographs and remote sensor data for practical evaluations of hazardous waste sites. *Photogrammetric Engineering and Remote Sensing* 53:515–519.

Lyon, J. 1993. *Practical Handbook for Wetlands Identification and Delineation*. Boca Raton, FL: CRC Press/Lewis Publishers.

Lyon, J. 2003. *Geographic Information System Applications for Watershed and Water Resources Management*. London: Taylor & Francis.

Lyon, J., and R. Drobney. 1984. Lake level effects as measured from aerial photos. *Journal of Surveying Engineering* 110:103–111.

Lyon, J., and R. G. Greene. 1992a. Use of aerial photographs to measure the historical areal extent of Lake Erie coastal wetlands. *Photogrammetric Engineering and Remote Sensing* 58:1355–1360.

Lyon, J., and R. Greene. 1992b. Lake Erie water level effects on wetlands as measured from aerial photographs. *Photogrammetric Engineering and Remote Sensing* 58:1355–1360.

Lyon, J., and L. Lyon. 2011. *Practical Handbook for Wetland Identification and Delineation*, 2nd ed. Boca Raton, FL: CRC Press.

Lyon, J., and J. McCarthy. 1981. SEASAT Radar Imagery for Detection of Coastal Wetlands. Ann Arbor, MI: 15th International Symposium on Remote Sensing of Environment.

Lyon, J., and J. McCarthy. 1995. *Wetland and Environmental Applications of GIS*. Boca Raton, FL: Lewis Publishers.

Lyon, J., R. Drobney, and C. Olson. 1986. Effects of Lake Michigan water levels on wetland soil chemistry and distribution of plants in the Straits of Mackinac. *Journal of Great Lakes Research* 12:175–183.

Lyon, J., K. Bedford, J. Yen Chien-Ching, D. Lee, and D. Mark. 1988. Suspended sediment concentrations as measured from multidate Landsat and AVHRR data. *Remote Sensing of Environment* 25:107–115.

Lyon, J., D. Yuan, R. Lunetta, and C. Elvidge. 1998. A change detection experiment using vegetation indices. *Photogrammetric Engineering and Remote Sensing* 64:143–150.

MacArthur, R., and E. Wilson. 1967. *The Theory of Island Biogeography*. Princeton, NJ: Princeton University Press.

Madden, M. 2015. *Landscape Analysis Using Geospatial Tools*. New York: Springer Publishing Company.

Maidment, D. 2002. *Arc Hydro: GIS for Water Resources*. Redlands, CA: ESRI Press.

Maidment, D., and D. Djokic. 2000. *Hydrologic and Hydraulic Modeling Support with GIS.* Redlands, CA: ESRI Press.

Mandelbrot, B. B. 1977. *Fractals, Form, Chance, and Dimension.* San Francisco: Freeman.

Marcaccio, J. V., and P. Chow-Fraser. 2014. Mapping options to track invasive phragmites Australis in the Great Lakes Basin in Canada in Proceedings of the 3rd International Conference—Water resources and wetlands. September 8–10, Tulcea (Romania); 75–82.

Mann, C. C. 2011. The dawn of the homogecene. *Orion Magazine* 30:16.

Marks, M., B. Lapin, and J. Randall. 1994. *Phragmites australis (P. communis)*: Threats, management, and monitoring. *Natural Areas Journal* 14:285–294.

Martz, L., and J. Garbrecht. 2003. Automated extraction of drainage network and watershed data from digital elevation models. *Water Resources Bulletin* 29:901–908.

Matkan, A., A. Shakiba, H. Pourali, and H. Azari. 2009. Flood early warning with integration of hydrologic and hydraulic models, remote sensing and GIS (case study: Madarsoo Basin, Iran). *World Applied Science Journal* 6:1698–1704.

May, C., R. Horner, J. Karr, B. Mar, and E. Welch. 1997. Effects of urbanization on small streams in the Puget Sound Lowland Ecoregion. *Watershed Protection Techniques* 2:483–493.

McDonnell, M. 1984. Interactions between Landscape Elements: Dispersal of bird disseminated plants in post agricultural landscapes, in *Methodology in Landscape Ecological Research and Planning*, eds. J. Brandt and R. Agger. Roskilde, Denmark: International Association of Landscape Ecologists, 47–58.

McDonnell, M., and E. Stiles. 1983. The structural complexity of old field vegetation and the recruitment of bird-dispersed plant species. *Oecologia* 56:109–116.

McEnroe, B. M., and P. Gonzalez. 2003. Storm Duration and Antecedent Moisture Conditions for Flood Discharge Estimation. Report No. K-TRAN: KU-02-4, Final Report, 1–50.

McGarigal, K. 2002. Landscape pattern metrics, in *Encyclopedia of Environmentrics*, volume 2, eds. A. El-Shaarawi, and W. Piegorsch. Sussex, UK: John Wiley & Sons.

McGarigal, K., and B. J. Marks. 1994. *Fragstats: Spatial Pattern Analysis Program for Quantifying Landscape Structure, Version 2.0.* Corvallis, OR: Oregon State University. https://www.umass.edu/landeco/pubs/mcgarigal.marks.1995.pdf.

McIntyre, N., and J. Wiens. 1999a. How does habitat patch size affect animal movement? An experiment with darkling beetles. *Ecology* 80:2262–2270.

McIntyre, N., and J. Wiens. 1999b. Interactions between habitat abundance and configuration: Experiment validation of some predictions from percolation theory. *Oikos* 86:129–137.

Mehaffey, M., M. Nash, T. Wade, D. Ebert, K. Jones, and A. Rager. 2005. Linking land cover and water quality in New York City's water supply watersheds. *Environmental Monitoring and Assessment* 107:29–44.

Meltzer, M. I., and H. M. Hastings. 1992. The use of fractals to assess the ecological impact of increased cattle population: Case study from the Runde Communal Land, Zimbabwe. *Journal of Applied Ecology* 29:635–646.

Messer, J. J., R. A. Linthurst, and W. S. Overton. 1991. EPA Program for monitoring ecological status and trends. *Environmental Monitoring and Assessment* 17:67–78.

Meyer, J. L., and G. E. Likens. 1979. Transport and transformation of phosphorus in a forest stream ecosystem. *Ecology* 60:1255–1269.

Meyer, S. R., K. Beard, C. S. Cronan, and R. J. Lilieholm. 2015. An analysis of spatio-temporal landscape patterns for protected areas in northern New England: 1900–2010. *Landscape Ecology* 30:1291–1305.

Millennium Ecosystem Assessment (MEA). 2005. *Ecosystems and Human Well-Being: Biodiversity Synthesis.* Washington, DC: World Resources Institute.

Miller, R. B., and C. Small. 2003. Cities from space: Potential applications of remote sensing in urban environmental research and policy. *Environmental Science Policy* 6:129–137.

Miller, W., and F. Egler. 1950. Vegetation of the Wequetequock-Paw-Catuck tidal marshes, Connecticut. *Ecological Monographs* 20:147–171.

Milly, P., J. Betancourt, M. Falkenmark, R. Hirsch, Z. Kundzewicz, D. Lettenmaier, and R. Stouffer. 2008. Stationary is dead: Whither water management? *Science* 319:573–574.

Milly, P., R. Wetherald, K. Dunne, and T. Delworth. 2002. Increasing risk of great floods in a changing climate *Nature* 415:514–517.

Milne, B. T. 1991. Lessons from applying fractal models to landscape patterns, in *Quantitative Methods in Landscape Ecology*, eds. M. G. Turner and R. H. Gardner. New York: Springer-Verlag, 199–235.

Missouri Resource Assessment Partnership. 2008. MoRAP webpage. Accessed from https://morap.missouri.edu/ (checked January 25, 2017). Columbia, MO: University of Missouri.

Mitsch, W., and S. Jorgensen. 2004. *Ecological Engineering and Ecosystem Restoration.* Hoboken, NJ: John Wiley & Sons.

Mitsch, W., J. Gosselink, L. Zhang, and C. Anderson. 2009. *Wetland Ecosystems.* New York: John Wiley & Sons.

Mitsch, W., J. Day, J. Gilliam, P. Groffman, D. Hey, G. Randall, and N. Wang. 2001. Reducing nitrogen loading to the Gulf of Mexico from the Mississippi river basin: Strategies to counter a persistent ecological problem. *Bioscience* 51:373–388.

Mladenoff, D. J., M. A. White, and J. Pastor. 1993. Comparing spatial pattern in unaltered old-growth and disturbed forest landscapes. *Ecological Applications* 3:294–306.

Moller, T., and C. Rordam. 1985. Species numbers of vascular plants in relation to area, isolation and age of ponds in Denmark. *Oikos* 45:8–16.

Molina, M. J., and F. S. Rowland. 1974. Stratospheric sink for chlorofluoromethands: Chlorine atom-catalysed destruction of ozone. *Nature* 249:810–812.

Moorhead, K., and M. Brinson. 1995. Response of wetlands to rising sea level in the Lower Coastal Plain of North Carolina. *Ecological Applications* 5:261–271.

Mudie, P., and R. Byearne. 1980. Pollen evidence for historic sedimentation rates in California coastal marshes. *Estuarine and Coastal Marine Science* 10:305–316.

Murkin, H., and X. Kale. 1986. Relationships between waterfowl and macroinvertebrate densities in a northern prairie marsh. *Journal of Wildlife Management* 50:212–217.

Musick, H. B., and H. D. Grover. 1991. Image textural measures as indices of landscape pattern, in *Quantitative Methods in Landscape Ecology*, eds. M. G. Turner and R. H. Gardner. New York: Springer-Verlag, 77–103.

Na, X., S. Zhang, X. Li, H. Yu, and C. Liu. 2010. Improved land cover mapping using random forests combined with Landsat Thematic Mapper imagery and ancillary geographic data. *Photogrammetric Engineering and Remote Sensing* 76:833–840.

Næsset, E. 2002. Predicting forest stand characteristics with airborne scanning laser using a practical two-stage procedure and field data. *Remote Sensing of Environment* 80:88–99.

Nagendraa, H., R. Lucasb, J. P. Honradoc, R. H. G. Jongmand, C. Tarantinoe, M. Adamoe, and P. Mairotaf. 2013. Remote sensing for conservation monitoring: Assessing protected areas, habitat extent, habitat condition, species diversity, and threats. *Ecological Indicators* 33:45–59.

Nagasaka, A., and F. Nakamura. 1999. The influences of land-use changes on hydrology and riparian environment in a northern Japanese landscape. *Landscape Ecology* 14:543–556.

Nash, M., D. Chaloud, and R. Lopez. 2005. Multivariate analyses (canonical correlation analysis and partial least Square, PLS) to model and assess the association of landscape metrics to surface water chemical and biological properties using Savannah River basin data. EPA/600/X-05/004. Washington DC: United States Environmental Protection Agency.

National Wetlands Inventory. 2012. NOAA Coastal Services Center digital coast web site and U.S. Fish and Wildlife Service. Accessed from https://coast.noaa.gov/dataregistry/search/collection/info/nwi or http://www.fws.gov/wetlands/ (checked January 25, 2017).

National Wetland Plants List. 2012. Accessed from http://geo.usace.army.mil/wetland_plants/index.html (checked January 25, 2017).

Niedzwiedz, W., and L. Ganske. 1991. Assessing lakeshore permit compliance using low altitude oblique 35-mm aerial photography. *Photogrammetric Engineering and Remote Sensing* 57:511–518.

NOAA, National Oceanic and Atmospheric Administration. 2007. DEM Development Report: Creation of the North Carolina Sea Level Rise Digital Elevation Model.

NOAA, National Oceanic and Atmospheric Administration. 2010. Tides and Currents, Historical Tide Data, Station Datums, and Monthly Means. NOAA Center for Operational Oceanographic Products and Services. Accessed from http://www.tidesandcurrents.noaa.gov (checked January 25, 2017).

Odum, E. 1985. Trends expected in stressed ecosystems. *Bioscience* 35:419–422.

Office of Science and Technology Policy. 2010. Achieving and sustaining earth observations: A preliminary plan based on a strategic assessment by the U.S. Group on Earth Observations. Washington DC: The White House.

Ogutu, Z. 1996. Multivariate analysis of plant communities in the Narok district, Kenya: The influence of environmental factors and human disturbance. *Vegetatio* 126:181–189.

O'Hara, C., T. Cary, and K. Schuckman. 2010. Integrated technologies for orthophoto accuracy verification and review. *Photogrammetric Engineering and Remote Sensing* 76:1097–1103.

Olivera, F., and D. R. Maidment. 2000. GIS tools for HMS modelling support, in *Hydrologic and Hydraulic Modelling with Geographic Information Systems*, eds. D. R. Maidment and D. Djokic. San Diego, CA: ESRI Press.

O'Neill, R. 2001. Predicting nutrient and sediment loadings to streams from landscape metrics: A multiple watershed study from the United States Mid-Atlantic Region. *Landscape Ecology* 16:301–312.

O'Neill, R., C. Hunsaker, and D. Levine. 1992. Monitoring challenges and innovative ideas, in *Ecological Indicators*, eds. D. McKenzie, D. Hyatt, and V. McDonald. London: Elsevier Applied Science, 1443–1460.

O'Neill, R. V., C. T. Hunsaker, S. P. Timmins, and B. L. Jackson. 1996. Scale problems in reporting landscape patterns at the regional scale. *Landscape Ecology* 11:169–180.

O'Neill, R. V., K. H. Riitters, J. D. Wickham, and K. B. Jones. 1999. Landscape pattern metrics and regional assessment. *Ecosystem Health* 5:225–233.

O'Neill, R. V., J. R. Krummel, R. H. Gardner, G. Sugihara, B. Jackson, D. L. DeAngelis, B. T. Milne, M. G. Turner, B. Zygmut, S. W. Christensen, V. H. Dale, and R. L. Graham. 1988. Indices of landscape pattern. *Landscape Ecology* 1:153–162.

Opdam, P. 1990. Understanding the Ecology of Populations in Fragmented Landscapes. Trondheim, Norway: Transactions of the 19th IUGB Congress.

Opdam, P., R. Apeldoorn, A. Schotman, and J. Kalkhoven. 1993. Population responses to landscape fragmentation, in *Landscape Ecology of a Stressed Environment*, eds. C. Vos, and P. Opdam. London: Chapman and Hall, 143–171.

Oregon Department of Environmental Quality. 2003. BIOFILTERS (Bioswales, Vegetative Buffers, & Constructed Wetlands) for Storm Water Discharge Pollution Removal. Accessed from http://www.deq.state.or.us/wq/stormwater /docs/nwr/biofilters.pdf (checked January 25, 2017).

The Pacific Institute. 2008. GIS data downloads, BFE Raster. Accessed on September 20, 2010, from the Pacific Institute website at http://www.pacinst.org/reports/sea _level_rise/data/index.htm (checked January 25, 2017).

Pantaleoni, E., R. Wynne, J. Galbraith, and J. Campbell. 2009. Mapping wetlands using ASTER data: A comparison between classification trees and logistic regression. *International Journal of Remote Sensing* 30:3423–3440.

Parker, D. J. 2000. *Floods*. London: Routledge.

Parry, M., O. Canziani, J. Palutikof, P. Van der Linden, and C. Hanson. 2007. *Climate Change 2007: Impacts, Adaption and Vulnerability*. Contribution of Working Group II to the Fourth Assessment Report of the Intergovernmental Panel on Climate Change (IPCC). New York: Cambridge University Press

Pastor, J., and M. Broschart. 1990. The spatial pattern of a northern conifer-hardwood landscape. *Landscape Ecology* 4:55–68.

Pastorok, R. A., S. M. Bartell, S. Ferson, and L. R. Ginzburg (eds.). 2016. *Ecological Modeling in Risk Assessment: Chemical Effects on Populations, Ecosystems, and Landscapes*. Boca Raton, FL: CRC Press.

Patel, D. P., and P. K. Srivastava. 2013. Flood hazards mitigation analysis using remote sensing and GIS: Correspondence with town planning scheme. *Water Resources Management* 27:2353–2368.

Peterjohn, W., and D. Corel. 1984. Nutrient dynamics in an agricultural watershed: Observations on the role of a riparian forest. *Ecology* 65:1466–1475.

Pickett, S., and J. Thompson. 1978. Patch dynamics and the design of natural reserves. *Biological Conservation* 13:27–37.

Plexida, S. G., A. I. Sfougaris, I. P. Ispikoudis, and V. P. Papanastasis. 2014. Selecting landscape metrics as indicators of spatial heterogeneity—A comparison among Greek landscapes. *International Journal of Applied Earth Observation and Geoinformation* 26:26–35.

Poiani, K., and P. Dixon. 1995. Seed banks of Carolina bays: Potential contributions from surrounding landscape vegetation. *American Midland Naturalist* 134:140–154.

Pollard, T., I. Eden, J. Mundy, and D. Cooper. 2010. A volumetric approach to change detection in satellite images. *Photogrammetric Engineering and Remote Sensing* 76:817–831.

Prince, H. H., P. J. Padding, and R. W. Knapton. 1992. Waterfowl use of the Laurentian Great Lakes. *Journal of Great Lakes Research* 18:673–699.

Raabe, E., and R. Stumpf. 1995. Monitoring tidal marshes of Florida's Big Bend. *Proceedings of the Third Thematic Conference on Remote Sensing of Marine and Coastal Environments* 2:483–494. Ann Arbor, MI: Environmental Research Institute of Michigan.

Radwell, A. J., and T. J. Kwak. 2005. Assessing ecological integrity of Ozark rivers to determine suitability for protective status. *Environmental Management* 35:799–810.

Rahmstorf, S. 2007. A semi-empirical approach to projecting future sea-level rise. *Science* 315:368–370.

Ramsey III, E., and J. R. Jensen. 1995. Modeling mangrove canopy reflectance using a light interaction model and an optimization technique. *Wetland and Environmental Applications of GIS*, 164–181.

Ramsey, E., G. Nelson, F. Baarnes, and R. Spell. 2004. Light attenuation profiling as an indicator of structural changes in coastal marshes, in *Remote Sensing and GIS Accuracy Assessment*, eds. R. Lunetta and J. Lyon. New York: CRC Press, 69–89.

Ramsey, E., A. Rangoonwala, Y. Suzuoki, and C. Jones. 2011a. Oil detection in a coastal marsh with polarimetric synthetic aperture radar (SAR). *Journal of Remote Sensing* 3:2630–2662.

Ramsey, E., D. Chappell, D. Jacobs, S. Sapkota, and D. Baldwin. 1998. Resource management of forested wetlands: Hurricane impact and recovery mapped by combining Landsat TM and NOAA AVHRR data. *Photogrammetric Engineering and Remote Sensing* 64:733–738.

Ramsey, E., Z. Lu, Y. Suzuoki, A. Rangoonwala, and D. Werle. 2011b. Monitoring duration and extent of storm surge flooding along the Louisiana coast with Envisat ASAR data. *IEEE Journal of Selected Topics on Applied Earth Observations and Remote Sensing* 4:387–399.

Ramsey, E., G. Nelson, S. Sapkota, S. Laine, J. Verdi, and S. Krasznay. 1999. Using multiple polarization L-band radar to monitor marsh burn recovery. *IEEE Transactions Geoscience and Remote Sensing* 37:635–639.

Rapport, D. J. 1990. Challenges in the detection and diagnosis of pathological change in aquatic ecosystems. *Journal of Great Lakes Research* 16:609–618.

Reddy, K. R., and R. D. DeLaune. 2008. *Biogeochemistry of Wetlands: Science and Applications*. CRC Press.

Reid, W. V., D. Chen, L. Goldfarb, H. Hackmann, Y. T. Lee, K. Mokhele, E. Ostrom, K. Raivio, J. Rockström, H. J. Schellnhuber, and A. Whyte. 2010. Earth system science for global sustainability: Grand challenges. *Science* 330:916–917.

Revell, D. L. 2007. Evaluation of Long-term and Storm Event Changes to the Beaches of the Santa Barbara Sandshed. Dissertation. Santa Cruz, CA: University of California.

Rex, K. D., and G. P. Malanson. 1990. The fractal shape of riparian forest patches. *Landscape Ecology* 4:249–258.

Riitters, K., and J. Coulston. 2005. Hot spots of perforated forest in the Eastern United States. *Environmental Management* 35:483–492.

Riitters, K. H., R. V. O'Neill, C. T. Hunsaker, J. D. Wickham, D. H. Yankee, S. P. Timmins, K. B. Jones, and B. L. Jackson. 1995. A factor analysis of landscape pattern and structure metrics. *Landscape Ecology* 10:23–39.

Riitters, K., J. Wickham, R. O'Neill, K. Jones, and E. Smith. 2000. Global-scale patterns of forest fragmentation. *Conservation Ecology* 4:3.

Rodrigues-Galiano, V., B. Ghimire, E. Pardo-Iguzquiza, M. Chica-Olmo, and R. Congalton. 2012. Incorporating the downscaled Landsat TM thermal band in land-cover classification using random forest. *Photogrammetric Engineering and Remote Sensing* 78:129–137.

Rogers, L., and A. Allen. 1987. Habitat Suitability Index Models: Black Bear, Upper Great Lakes Region. Biological Report 82(10.144). Washington, DC: Department of Interior, U.S. Fish and Wildlife Service.

Roth, N., J. Allan, and D. Erickson. 1996. Landscape influences on stream biotic integrity assessed at multiple spatial scales. *Landscape Ecology* 11:141–156.

Running, S. W., R. R. Nemani, and R. D. Hungerford. 1987. Extrapolation of synoptic meteorological data in mountainous terrain and its use for simulating forest evapotranspiration and photosynthesis. *Canadian Journal of Forest Research* 17:472–483.

Running, S. W., R. R. Nemani, F. A. Heinsch, M. Zhao, M. Reeves, and H. Hashimoto. 2004. A continuous satellite-derived measure of global terrestrial primary production. *AIBS Bulletin* 54(6):547–560.

Sadro, S., M. Gastil-Buhl, and J. Melack. 2007. Characterizing patterns of plant distribution in a southern California salt marsh using remotely sensed topographic and hyperspectral data and local tidal fluctuations. *Remote Sensing and the Environment* 110:226–239.

Sarkar, S., S. M. Parihar, and A. Dutta. 2016. Fuzzy risk assessment modelling of East Kolkata Wetland Area: A remote sensing and GIS based approach. *Environmental Modelling and Software* 75:105–118.

Schaal, G. 1995. *Methods Used in the Ohio Wetland Inventory*. Columbus: OH: Ohio Department of Natural Resources.

Schlesinger, W., and E. S. Bernhardt. 2013. Biogeochemistry: An Analysis of Global Change. Oxford, UK: Elsevier.

Schlesinger, W. H., M. C. Dietze, R. B. Jackson, R. P. Phillips, C. C. Rhoades, L. E. Rustad, and J. M. Vose. 2015. Forest biogeochemistry in response to drought. *Global Change Biology* 22:2318–2328.

Schott, J. 2007. *Remote Sensing: The Image Chain Approach*. New York: Oxford University Press.

Scott, J., F. Davis, B. Csulti, R. Noss, B. Butterfield, C. Groves, H. Anderson, S. Caicco, F. D'Erchia, T. Edwards, J. Ulliman, and R. Wright. 1993. Gap analysis: A geographic approach to protection of biological diversity. *Wildlife Monographs* 123:1–41.

Semadeni-Davies, A., S. Elliott, and U. Shankar. 2014. How CLUES can help in managing catchment nutrient limits: Nutrient management for the farm, catchment and community. Occasional Report 27.

Sheaves, M., R. Baker, I. Nagelkerken, and R. M. Connolly. 2015. True value of estuarine and coastal nurseries for fish: Incorporating complexity and dynamics. *Estuaries and Coasts* 38:401–414.

Shuman, C., and R. Ambrose. 2003. A comparison of remote sensing and ground-based methods for monitoring wetland restoration success. *Restoration Ecology* 11:325–333.

Simberloff, D., and L. Abele. 1982. Refuge design and island biogeographic theory: Effects of fragmentation. *The American Naturalist* 120:41–50.

Simberloff, D., and E. Wilson. 1970. Experimental zoogeography of islands: A two-year record of colonization. *Ecology* 51:934–937.

Singh, A., 1989. Digital change detection techniques using remotely-sensed data. *International Journal of Remote Sensing* 10:989–1003.

Sinha, P., L. Kumar, and N. Reid. 2012. Seasonal variation in land-cover classification accuracy in a diverse region. *Photogrammetric Engineering and Remote Sensing* 78:271–280.

Schmidt, M., R. Lucas, P. Bunting, J. Verbesselt, and J. Armston. 2015. Multi-resolution time series imagery for forest disturbance and regrowth monitoring in Queensland, Australia. *Remote Sensing of Environment* 158:156–168.

Smith, S. D., P. B. McIntyre, B. S. Halpern, R. M. Cooke, A. L. Marino, G. L. Boyer, A. Buchsbaum, G. A. Burton, L. M. Campbell, J. J. Ciborowski, and P. J. Doran. 2015. Rating impacts in a multi-stressor world: A quantitative assessment of 50 stressors affecting the Great Lakes. *Ecological Applications* 25:717–728.

SOLEC. 2000. Selection of Indicators for Great Lakes Basin Ecosystem Health. U.S. Environmental Protection Agency.

SOLEC. 2009. State of the Great Lakes 2009 (EPA 905-R-09-031). Environment Canada and United States Environmental Protection Agency.

Soule, M., B. Wilcox, and C. Holtby. 1979. Benign neglect: A model of faunal collapse in game reserves of East Africa. *Biological Conservation* 15:259–272.

Stevens, D., and S. Jensen. 2007. Sample design, execution, and analysis for wetland assessment. *Wetlands* 27:515–523.

Stiling, P. 1996. *Ecology: Theories and Applications*, 2nd ed. Upper Saddle River, NJ: Prentice Hall.

Stone, R. 2010. Earth-observation summit endorses global data sharing. *Science* 330:902.

Stuckey, R. 1989. Western Lake Erie aquatic and wetland vascular plant flora: Its origin and change, in *Lake Erie Estuarine Systems: Issues, Resources, Status, and Management*, ed. K. Krieger. Washington, DC: Department of Commerce, National Oceanographic and Atmospheric Administration.

Sugihara, G., and R. M. May. 1990. Applications of fractals in ecology. *Trends in Ecology and Evolution* 5:79–86.

Sun, D., Y. Yu, R. Zhang, S. Li, and M. Goldberg. 2012. Towards operational automatic flood detection using EOS/MODIS data. *Photogrammetric Engineering and Remote Sensing* 78:637–646.

Tang, J., P. V. Bolstad, and J. G. Martin. 2009. Soil carbon fluxes and stocks in a Great Lakes forest chronosequence. *Global Change Biology* 15:145–155.

Thenkabail, P., J. Lyon, and A. Huerte. 2012. *Hyperspectral Remote Sensing of Vegetation.* Boca Raton, FL: CRC Press.

Thenkabail, P., M. Schull, and H. Turral. 2005. Ganges and Indus river basin land use/land cover (LULC) and irrigated area mapping using continuous streams of MODIS data. *Remote Sensing Environment* 95:317–341.

Thenkabail, P., J. Lyon, H. Turral, and C. Biradar. 2009. *Remote Sensing of Global Croplands for Food Security.* Boca Raton, FL: CRC Press.

Tiner, R. 1999. *Wetland Indicators: A Guide to Wetland Identification, Delineation.* Boca Raton, FL: CRC Press.

Tiner, R., H. Bergquist, G. DeAlessio, and M. Starr. 2002. *Geographically Isolated Wetlands: A Preliminary Assessment of their Characteristics and Status in Selected Areas of the U.S.* Hadley, MA: Department of the Interior, U.S. Fish and Wildlife Service.

Touzi, R., A. Deschamps, and G. Rother. 2007. Wetland characterization using polarimetric Radarsat-2 capability. *Canadian Journal of Remote Sensing* 33:S56–S67.

Turner, M. G. 1989. Landscape ecology: The effect of pattern on process. *Annual Review of Ecology and Systematics* 20:171–197.

Turner, M. G. 1990a. Spatial and temporal analysis of landscape pattern. *Landscape Ecology* 3:153–162.

Turner, M. G. 1990b. Landscape changes in nine rural counties of Georgia. *Photogrammetric Engineering and Remote Sensing* 56:379–386.

Turner, M. G., and R. H. Gardner (eds.). 1991. *Quantitative Methods in Landscape Ecology.* New York: Springer-Verlag.

Turner, M. G., and C. L. Ruscher. 1988. Changes in landscape patterns in Georgia, USA. *Landscape Ecology* 1:241–251.

Turner, M. G., R. H. Gardner, and R. V. O'Neill. 2001. *Landscape Ecology in Theory and Practice.* New York: Springer-Verlag.

Turner, R. E., and N. N. Rabalais. 2004. Suspended sediment, C, N, P, and Si yields from the Mississippi River Basin. *Hydrobiologia* 511:79–89.

Twedt, D., and C. Loesch. 1999. Forest area and distribution in the Mississippi alluvial valley: Implications from breeding bird conservation. *Journal of Biogeography* 26:1215–1224.

UN (United Nations). 2010. World Urbanization Prospects, the 2009 Revision. United Nations, Department of Economic and Social Affairs, Population Division.

UN (United Nations). 2012. Sustainable Development in the 21st Century: Sustainable Land Use for the 21st Century (SD21). Department of Economic and Social Affairs.

U.S. Army Corps of Engineers. 1987. Corps of Engineers Wetlands Delineation Manual. Technical Report Y-87-1. Vicksburg, MS: Environmental Laboratory, U.S. Army Engineer Waterways Experiment Station.

U.S. Army Corps of Engineers. 1993. Photogrammetric Mapping. Engineering Manual. Washington, DC.

U.S. Census Bureau. 2010. North Carolina Data. Retrieved from https://www.census .gov/newsroom/releases/archives/2010_census/cb11-cn159.html (checked July 7, 2017).

USDA. 1986. Technical Release 55, Urban Hydrology for Small Watersheds. U.S. Department of Agriculture, Natural Resource Conservation Service, Washington, DC.

USDA. 2001. Summary Report: 1997 National Resources Inventory (revised December 2001). U.S. Department of Agriculture, Natural Resource Conservation Service, Washington, DC.

USDA and USEPA. 1994. *A Handbook for Constructed Wetlands: A Guide to Creating Wetlands for Agricultural Wastewater, Domestic Wastewater, Coal Mine Drainage and Stormwater in the Mid-Atlantic Region. Vol. 1: General Considerations.* National Environmental Publications Information System. Accessed from http://nepis .epa.gov (checked January 25, 2017).

USEPA (U.S. Environmental Protection Agency). 1991. Federal Manual for Identifying and Delineating Jurisdictional Wetlands. Federal Register August 14.

USEPA (U.S. Environmental Protection Agency). 1994. Landscape Monitoring and Assessment Research Plan, EPA 620/R–94/009. Office of Research and Development, Washington, DC.

USEPA (U.S. Environmental Protection Agency). 1995. Ecological Restoration: A Tool to Manage Stream Quality, EPA/841/F-95/007. USEPA, Washington, DC.

USEPA (U.S. Environmental Protection Agency). 1996. Mid-Atlantic Landscape Indicators Project Plan, EPA 620/DRAFT. Office of Research and Development, Washington, DC.

USEPA (U.S. Environmental Protection Agency). 1997. Monitoring Water Quality: Volunteer Stream Monitoring—A Methods Manual, EPA/841/B-97/003. USEPA, Washington, DC.

USEPA (U.S. Environmental Protection Agency). 2001a. Landscape Analysis and Assessment—Overview. U.S. Environmental Protection Agency, National Exposure Research Laboratory, Environmental Sciences Division, Las Vegas, NV.

USEPA (U.S. Environmental Protection Agency). 2001b. National Coastal Condition Report, EPA-620/R-01/005. U.S. Environmental Protection Agency, Office of Research and Development, Office of Water, Washington, DC.

USEPA (U.S. Environmental Protection Agency). 2002. Great Lakes Coastal Wetlands: Abiotic and Floristic Characterization. U.S. Environmental Protection Agency, Great Lakes National Program Office, Chicago, IL. Accessed from http://www.epa.gov/glnpo/ecopage/wetlands/glc/glctext.html (checked January 25, 2017).

USEPA (U.S. Environmental Protection Agency). 2003. Protecting Wetlands along the Great Lakes Shoreline. U.S. Environmental Protection Agency, Washington, DC.

USEPA (U.S. Environmental Protection Agency). 2004a. Great Lakes Fact Sheet. U.S. Environmental Protection Agency, Great Lakes National Program Office, Chicago, IL.

USEPA (U.S. Environmental Protection Agency). 2004b. ATtILA User guide. EPA/600/R-04/083. USEPA Washington, DC.

USEPA (U.S. Environmental Protection Agency). 2008. STOrage and RETrieval—EPA's main Repository for Water Quality, Biological and Physical Data. USEPA, Washington, DC. Accessed from http://www.epa.gov/storet/index.html (checked January 25, 2017).

USGS. 2003. United States Geologic Survey, Effects of Urban Development on Floods. Accessed from http://pubs.usgs.gov/fs/fs07603/pdf/fs07603.pdf (checked January 25, 2017). USGS, Washington, DC.

USGS. 2005. Hydrologic and Water-Quality Conditions in the Kansas River, Northeast Kansas, November 2001–August 2002, and Simulation of Ammonia Assimilative Capacity and Bacteria Transport During Low Flow. United States Geological Survey.

USGS. 2006. United States Geologic Survey, Flood Hazards—A National Threat. Accessed from http://pubs.usgs.gov/fs/2006/3026/2006-3026.pdf (checked January 25, 2017). USGS, Washington, DC.

USGS. 2008. U.S. Geological Survey Surface: Water Data for the Nation. USGS, Washington, DC. Accessed from http://nwis.waterdata.usgs.gov/nwis/sw (checked January 25, 2017).

Van der Valk, A. 1981. Succession in wetlands: A Gleasonian approach. *Ecology* 62:688–696.

Van der Valk, A., and C. Davis. 1980. The impact of a natural drawdown on the growth of four emergent species in a prairie glacial marsh. *Aquatic Botany* 9:301–322.

Van Derventer, A. 1992. Evaluating the Usefulness of Landsat Thematic Mapper to Determine Soil Properties, Management Practices, and Soil Water Content. Doctoral dissertation. Columbus, OH: Ohio State University.

Van Remortel, R., R. Maichle, and R. Hickey. 2004. Computing the RUSLE LS factor through array-based slope length processing of digital elevation data using a C++ executable. *Computers and Geosciences* 30:1043–1053.

Van Remortel, R., R. Maichle, D. Huggem, and A. Pitchford. 2005. Automated GIS Watershed Analysis Tools for RUSLE/SEDMOD Soil Erosion and Sedimentation Modeling. EPA/600/X-05/007. Washington, DC: United States Environmental Protection Agency.

Van Sickle, J. 2008. *GPS for Land Surveyors*, 3rd ed. Boca Raton, FL: CRC Press.

Velpuri, N., P. Thenkabail, C. Gumma, C. Biradar, V. Dheeravath, P. Noojipady, and L. Yuanjie. 2009. Influence of resolution inn irrigated area mapping and area estimation. *Photogrammetric Engineering and Remote Sensing* 75:1383–1395.

Vermeer, M., and S. Rahmstorf. 2009. Global Sea Level Linked to Global Temperature. Accessed July 2010 from the Proceedings of the National Academy of Sciences website at www.pnas.org_cgi_doi_10.1073_pnas.0907765106.

Vernberg, W. B., and B. C. Coull. 1981. Meiofauna, in *Functional Adaptations of Marine Organisms*, eds. F. J. Vernberg and W. B. Vernberg. New York: Academic Press, 147–172.

Vicente-Serrano, S., F. Perez-Cabello, and T. Lasanta. 2008. Assessment of radiometric correction techniques in analyzing vegetation variability and change using time series of Landsat images. *Remote Sensing of Environment* 112:3916–3934.

Villeneuve, J. 2005. Delineating Wetlands Using Geographic Information System and Remote Sensing Technologies. Master's thesis. College Station, TX: Texas A&M University.

Vitousek, P. M., and J. M. Melillo. 1979. Nitrate losses from disturbed forests: Patterns and mechanisms. *Forest Science* 25:605–619.

Vogelmann, J., D. Helder, R. Morfitt, R. M. Choate, J. Merchant, and H. Bulley. 2010. Effects of Landsat 5 Thematic Mapper and Landsat 7 Enhanced Thematic Mapper Plus radiometric and geometric calibrations and corrections on landscape. *Hydrology and Earth System Sciences* 14:2415–2428.

Vogelmann, J. E., S. M. Howard, L. Yang, C. R. Larson, B. K. Wylie, and N. Van Driel. 2001. Completion of the 1990s National Land Cover Data Set for the conterminous United States from Landsat Thematic Mapper data and ancillary data sources. *Photogrammetric Engineering and Remote Sensing* 67:650–662.

Voss, C. M., R. R. Christian, and J. T. Morris. 2013. Marsh macrophyte responses to inundation anticipate impacts of sea-level rise and indicate ongoing drowning of North Carolina marshes. *Marine Biology* 160:81–194.

Wamsley, T. V., M. A. Cialone, J. M. Smith, J. H. Atkinson, and J. D. Rosati. 2010. The potential of wetlands in reducing storm surge. *Ocean Engineering* 37:59–68.

Wang, W. Y., G. L. Tang, Y. Zhou, and Q. Zhang. 2016. The study on ecological risk assessment in the port of the coastal city based on the Landscape Pattern Analysis, in *Civil Engineering and Urban Planning IV: Proceedings of the 4th International Conference on Civil Engineering and Urban Planning*. Beijing, China: CRC Press.

Ward, A., and S. Trimble. 2003. *Environmental Hydrology*, 2nd ed. Boca Raton, FL: CRC Press.

WDR. 2010. World Disasters Report 2010. International Federation of Red Cross and Red Crescent Societies.

Weiss, D., J. Callaway, and R. Gersberg. 2001. Vertical accretion rates and heavy metal chronologies in wetland sediments of the Tijuana Estuary. *Estuaries* 24:840–850.

Whigham, D., D. Weller, A. Jacobs, T. Jordan, and M. Kentula. 2003. Assessing the ecological condition of wetlands at the catchment scale. *Landscape Ecology* 20:99–111.

Whiteaker, T. L., O. Robayo, D. R. Maidment, and D. Obenour. 2006. From a NEXRAD rainfall map to a flood inundation map. *Journal of Hydrolic Engineering* 11:37–45.

Wickham, J. D., and K. H. Riitters. 1995. Sensitivity of landscape metrics to pixel size. *International Journal of Remote Sensing* 16:3585–3594.

Wickham, J. D., K. H. Riitters, R. V. O'Neill, K. B. Jones, and T. D. Wade. 1996. Landscape "contagion" in raster and vector environments. *International Journal of Remote Sensing* 10:891–899.

Wilcox, D. 1995. Wetland and aquatic macrophytes as indicators of anthropogenic hydrologic disturbance. *Natural Areas Journal* 15:240–248.

Williams, D., and J. Lyon. 1991. Use of a geographical information system data base to measure and evaluate wetland changes in the St. Marys River, Michigan. *Hydrobiologia* 219:83–95.

Willis, C., and W. Mitsch. 1995. Effects of hydrology and nutrients on seedling emergence and biomass of aquatic macrophytes from natural and artificial seed banks. *Ecological Engineering* 4:65–76.

Wilson, E. O. (ed.). 1988. *Biodiversity.* Washington, DC: National Academy Press.

Wold, S. 1995. PLS for multivariate linear modeling, in *Chemometric Methods in Molecular Design Methods and Principles in Medicinal Chemistry,* ed. H. Van de Waterbeemd. Weinheim, Germany: Verlag-Chemie, 195–218.

Wolter, P., C. Johnston, and G. Niemi, 2005. Mapping submerged aquatic vegetation in the U.S. Great Lakes using QuickBird satellite data. *International Journal of Remote Sensing* 26:5255–5274.

Wu, S. 1989. Multipolarization P-, L-, and C-Band Radar for Coastal Zone Mapping: The Louisiana Example. Cleveland, OH: The 1989 Fall Convention of ACSM/ASPRS.

Wyrick, J. R., B. A. Rischman, C. A. Burke, C. McGee, and C. Williams. 2009. Using hydraulic modeling to address social impacts of small dam removals in southern New Jersey. *Journal of Environmental Management* 90(Supplement 3): S270–S278.

Yang, J., and F. Artigas. 2010. Mapping Salt Marsh Vegetation by Integrating Hyperspectral and LiDAR Remote Sensing, in *Remote Sensing of Coastal Environment,* ed. J. Wang. Boca Raton, FL: CRC Press, 173–190.

Yang, C., J. Everitt, R. Fletcher, J. Jensen, and P. Mausel. 2009. Mapping black mangrove along the south Texas gulf coast using AISA+ hyperspectral imagery. *Photogrammetric Engineering and Remote Sensing* 75:425–436.

Yi, G., D. Risley, M. Koneff, and C. Davis. 1994. Development of Ohio's GIS-based wetlands inventory. *Journal of Soil and Water Conservation* 49:23–28.

Young, S., and C. Wang. 2001. Land-cover change analysis of China using global-scale Pathfinder AVHRR Land cover (PAL) data, 1982–92. *International Journal of Remote Sensing* 22:1457–1477.

Zandbergen, P., and F. Petersen. 1995 (May 4). The Role of Scientific Information in Policy and Decision-Making: The Lower Fraser Basin in Transition. Symposium and Workshop. Kwantlen College, Surrey, British Columbia, Canada.

Zomer, R., A. Trabucco, and S. Ustin. 2009. Building spectral libraries for wetlands land cover classification and hyperspectral remote sensing. *Journal of Environmental Management* 90:2170–2177.

Metadata Cited

Data Title: Canada Hydrologic Units (Subsubdivisions)

Description: Hydrologic subsubdivisions comprising the Ontario Great Lakes. 1:2M-scale subsub-basins for processing and reporting on a suite of Great Lakes landscape metrics.

Originator: Geomatics for Sustainable Development of Natural Resources, Environment Canada, Ottawa, Ontario

Data Title: Nation Landcover Database

Description: Reference Homet et al. 2015 or www.mrlc.gov/publications.

Originator: U.S. Geological Survey

Data Title: NHDPlus, Edition 1.0

Description: The NHDPlus Version 1.0 is an integrated suite of application-ready geospatial data sets that incorporate many of the best features of the National Hydrography Dataset (NHD) and the National Elevation Dataset (NED). The NHDPlus includes a stream network (based on the 1:100,000-scale NHD), improved networking, naming, and "value-added attributes" (VAAs). NHDPlus also includes elevation-derived catchments (drainage areas) produced using a drainage enforcement technique first broadly applied in New England and thus dubbed "the New England Method." This technique involves "burning-in" the 1:100,000-scale NHD and when available building "walls" using the national Watershed Boundary Dataset (WBD). The resulting modified digital elevation model (HydroDEM) is used to produce hydrologic derivatives that agree with the NHD and WBD. An interdisciplinary team from the U.S. Geological Survey (USGS), U.S. Environmental Protection Agency (USEPA), and contractors, over the last two years has found this method to produce the best quality NHD catchments using an automated process.

Originators: U.S. Environmental Protection Agency and the U.S. Geological Survey

Data Title: Ontario Ministry of Natural Resources Data Set

Description: Ontario's land cover map reclassed to NOAA's Coastal Change Analysis Program (C-CAP) land cover classification. C-CAP includes the classification of 2000 Landsat 7 data to produce a land cover product intended to improve the understanding of coastal uplands and wetlands, and their linkages with the distribution, abundance, and health of living marine resources.

Originator: Ontario Ministry of Natural Resources, 300 Water Street, Peterborough, Ontario K9J 8M5

Data Title: United States Census 2000 (c2k)

Description: Census 2000 (c2k) census block groups for the Great Lakes Basin. Census block groups and corresponding centroid-based attribute data extracted from Census2000 blocks.

Originator: U.S. Bureau of Census, Washington, DC

Data Title: United States Coastal Change Analysis Program (CCAP)

Description: NOAA's Coastal Change Analysis Program (C-CAP) land cover geodata set. C-CAP includes the classification of 2000 Landsat 7 data to produce a land cover product intended to improve the understanding of coastal uplands and wetlands, and their linkages with the distribution, abundance, and health of living marine resources.

Originator: National Oceanic and Atmospheric Administration Coastal Services Center, Charleston, South Carolina, (843)740-1210, csc@csc .noaa.gov

Data Title: United States Digital General Soil Map

Description: This data set consists of general soil association units. It was developed by the National Cooperative Soil Survey and supersedes the State Soil Geographic (STATSGO) data set published in 1994. It consists of a broad-based inventory of soils and nonsoil areas that occur in a repeatable pattern on the landscape and that can be cartographically shown at the scale mapped. The data set was created by generalizing more detailed soil survey maps. Where more detailed soil survey maps were not available, data on geology, topography, vegetation, and climate were assembled, together with Land Remote Sensing Satellite (LANDSAT) images. Soils of like areas were studied, and the probable classification and extent of the soils were determined.

Map unit composition was determined by transecting or sampling areas on the more detailed maps and expanding the data statistically to characterize the whole map unit.

This data set consists of georeferenced vector digital data and tabular digital data. The map data were collected in 1- by 2-degree topographic quadrangle units and merged into a seamless national data set. It is distributed in state/territory and national extents. The soil map units are linked to attributes in the National Soil Information

System database, which gives the proportionate extent of the component soils and their properties.

Originator: U.S. Department of Agriculture, Natural Resources Conservation Service

Data Title: United States Hydrologic Units (8-digit HUCs)

Description: Catalog units (8-digit HUCs). 1:250K-scale 8-digit HUC subbasins for processing and reporting on the suite of landscape metrics.

Originator: U.S. Geological Survey, Reston, Virginia

Data Title: United States Roads (Wessex/GDT)

Description: Wessex roads (interstate highways, major roads, and streets) as derived from Bureau of the Census TIGER/Line files at 1:24000 scale or smaller. Highways, roads, and streets derived from the Wessex TIGER/Line.

Originators: Wessex (A Division of GDT), 11 Lafayette St., Lebanon, NH 03766-1455, 1 (800) 331-7881 and U.S. Department of Commerce, Bureau of the Census, Geography Division, Washington, DC, tiger@census.gov

Index

Page numbers followed by f and t indicate figures and tables, respectively.

256

Index